A Company in History
Solvay, 1863–2013

To what extent is a family business the outcome of a business family? Are inventors working on their own or relying on a wide array of associates? What is the impact of world politics on multinational companies? Do they "think global and act local" or do they proceed the other way around? This book intends to address these and other questions by shedding light on the history of the chemical company Solvay through 150 years of historical change and turmoil. The Solvay company was created in 1863 as a start-up enterprise manufacturing soda ash with a new industrial process. Its main founder, Ernest Solvay, was in his early twenties and had no university training in chemistry. Against all odds, the business rapidly grew out as one of the world's largest multinational chemical companies. In this book, the Solvay family business becomes a window enabling readers to see the unfolding of a century and a half of world history.

Kenneth Bertrams is Research Associate at the F.R.S. – Fonds National de la Recherche Scientifique and Lecturer at the Université Libre de Bruxelles, Belgium.

A Company in History
Solvay, 1863–2013

KENNETH BERTRAMS

Fonds National de la Recherche Scientifique, Belgium
Université Libre de Bruxelles

CAMBRIDGE
UNIVERSITY PRESS

CAMBRIDGE UNIVERSITY PRESS
Cambridge, New York, Melbourne, Madrid, Cape Town,
Singapore, São Paulo, Delhi, Mexico City

Cambridge University Press
32 Avenue of the Americas, New York, NY 10013-2473, USA

www.cambridge.org
Information on this title: www.cambridge.org/9781107607576

© Kenneth Bertrams 2013

This publication is in copyright. Subject to statutory exception
and to the provisions of relevant collective licensing agreements,
no reproduction of any part may take place without the written
permission of Cambridge University Press.

First published 2013

Printed in the United Kingdom at CPI

A catalog record for this publication is available from the British Library.

Library of Congress Cataloging in Publication Data
Bertrams, Kenneth.
A company in history : Solvay, 1863–2013 / Kenneth Bertrams, F.R.S. – Fonds
National de la Recherche Scientifique, Belgium, Université Libre de Bruxelles.
pages cm
Includes bibliographical references.
ISBN 978-1-107-60757-6 (pbk.)
1. Solvay Chemicals – History. 2. Chemical industry – Belgium – History
3. International business enterprises – Belgium – History. 4. Solvay, Ernest,
1838–1922. I. Title.
HD9656.B44S652   2013
338.8′8766009493–dc23   2012040595

ISBN 978-1-107-60757-6 Paperback

Cambridge University Press has no responsibility for the persistence or accuracy of
URLs for external or third-party Internet Web sites referred to in this publication
and does not guarantee that any content on such Web sites is, or will remain,
accurate or appropriate.

Parts of this book are based on the scholarly study published simultaneously:
Kenneth Bertrams, Nicolas Coupain, and Ernst Homburg, *Solvay: History of a
Multinational Family Firm*, Cambridge: Cambridge University Press, 2013,
630 pages, ISBN 978-1-107-02480-9.

Not for resale or exchange

CONTENTS

List of Figures	*page* ix
Foreword by Daniel Janssen	xiii
INTRODUCTION	1
1 THE GOLDEN AGE OF PROGRESS	5
Opening: Doctor Ferguson, Mister Solvay	5
Setting the Stage: The Triumph of Industrial Capitalism	7
The Burning Sun of Knowledge and the Art of Not Knowing	10
Merry Inventors, Reluctant Entrepreneurs	13
Self-Made Men Do Work in a Collective Environment	17
Soda Ash and the Ghost of Nicolas Leblanc	21
2 OUTLINE OF A FAMILY BUSINESS	25
Criteria and Commonplaces	25
The First and Closest Circle – The Solvay Clan	28
The Second and Third Circles: Relatives and Friends	33
Life in a Partnership Company	37
Outsiders Needed (or How to Avoid King Lear Syndrome)	40

Contents

3 **BUILDING AN INDUSTRIAL EMPIRE** — 45

"Faster, Higher, Stronger": Picturing the First Globalization — 45
How to Become French (in France) and German (in Germany) — 50
Stretching over the Channel — 55
Europe Is Not Enough — 58
A Multilayered Hegemony — 61

4 **WORLD WAR I AND THE COLLAPSE OF THE INTERNATIONAL ORDER** — 65

A Dive into the Dark — 65
"Never Was So Much Owed by So Many to So Few" — 67
Organizing the Economic Mobilization — 71
Into the War Economy — 72
The Grim Postwar (or the Pursuit of War by Other Means) — 75
Recasting Nations, Resuming Industrial Relations — 77
The Twilight of an Era — 81

5 **THE RATIONALIZATION OF THE WORLD CHEMICAL INDUSTRY** — 85

Bigger Businesses, More Products: A View from the United States — 85
Allied Chemical or the Mysterious Mister Weber — 88
The Synthetic Ammonia Years (1919–1924) — 93
The "Magic Square" Venture That Never Came Through (1925–1926) — 96
ICI, the British Empire of Chemistry — 99
Speaking of Crisis: Solvay & Cie at the End of the 1920s — 101

6 **FROM CRISIS TO WAR** — 107

Witnessing the Agony of Capitalism — 107
Recasting Industrial Stability — 110
Italy and Germany: Laboratories of Fascism — 112
1936: The Spanish Prelude — 117
Hitler's Greater Germany — 120

	The Economics of Occupation	123
	Meanwhile, across the Ocean...	126
7	RECONSTRUCTION THROUGH DIVERSIFICATION	129
	The Resettlement of Postwar Germany	129
	The Division of Europe and Its Harsh Consequences	134
	Western Europe: Recovery, Reconstruction, Integration	137
	The American Leadership between Constraints and Seduction	139
	Embarking on the Plastic Drive	144
	Diversifying the Diversification	146
8	RECESSION AND THE BIOCHEMICAL IMPULSE	151
	The End of the "Golden Age"	151
	First Strategy, Then Structure: Solvay & Cie Becomes a Public Company (1967)	154
	An American Comeback	158
	Caught Up by the Crisis	162
	Biochemical Innovation	165
9	GLOBALIZATION AND CONSOLIDATION	171
	The Cold War Is Over	171
	"The Wind of Change": Back to Central Eastern Europe	173
	Tigers and Dragons: Solvay in Asia	176
	After 1993, Focusing on the Core Business	180
	Looking Forward in the Twenty-First Century	184

LIST OF FIGURES

1.1	Machinery in Motion court, International Exhibition, London, 1862.	page 4
1.2	Louis-Philippe Acheroy in his workshop at Couillet.	12
1.3	Chemical works at Floreffe (Belgium) in about 1850.	15
1.4	Thomas Edison in his lab in Menlo Park, New Jersey.	19
1.5	Semi-industrial laboratory at Saint-Gilles (Brussels), around 1900.	22
2.1	Four generations of Solvay leaders.	24
2.2	Four Solvay presidents.	29
2.3	Porphyry quarries of Quenast.	30
2.4	Solvay managers visiting Wyhlen soda ash plant in Germany, around 1890.	36
2.5	Dinner hosted by Carl Wessel (Deutsche Solvay Werke manager).	42
3.1	Pumping engine at Solvayhall plant, Germany.	44
3.2	Bags of sodium carbonate (soda ash) at Rosignano plant, Italy.	48
3.3	Workers at the Bernburg plant, 1903.	54
3.4	Managers of Solvay & Cie and Brunner, Mond & Co visiting Syracuse plant, New York, around 1897.	60
3.5	Map of Solvay plants in 1913 (including subsidiaries and associated companies).	62

List of Figures

4.1 British soldiers digging a trench while wearing respirators to guard against fumes from bursting shells. 64
4.2 General meeting of the Comité National de Secours et d'Alimentation. 69
4.3 Results of the bombing of Château-Salins soda ash plant on 24 July 1917. 73
4.4 Solvay executives visiting Wieliczka salt mine, near Krakow, Poland (1921). 80
4.5 First Solvay Conference on Physics in Brussels, 1911. 82
5.1 Solvay Mercury cells at Jemeppe electrolytic plant, 1910. 84
5.2 The Allied Chemical building in New York, 1963. 90
5.3 Orlando Weber, the "mystery man" of Wall Street. 92
5.4 Advertisement for Imperial Chemical Industries. 100
5.5 Aerial view of Tavaux plant, soon after its construction in 1932. 103
6.1 Unemployed workers from the English town of Jarrow during their "Hunger march" to London in October 1936. 106
6.2 Map of Solvay plants in 1938. 113
6.3 A folk float decorated with fascist symbols, during the Festa dell'Uva in the village of Rosignano Solvay (1933). 116
6.4 Workers digging potash in the Suria mine in Barcelona's hinterland. 119
6.5 Vienna under Nazi rule after the Anschluss. 122
6.6 Ernest-John Solvay and René Boël. 125
7.1 Plastic bottles produced by Solvay. 128
7.2 The Potsdam Conference in Germany (July 1945). 132
7.3 The "Solvay Trial" at Bernburg, 14 December 1950. 136
7.4 Marshall Plan Funds provided $1,390,600 to the Federal Republic of Germany. 140
7.5 Vinyl records made of PVC symbolize entrance into the culture of mass consumption and the plastics era. 143
8.1 Drug manufacturing at Kali-Chemie (early 1990s). 150
8.2 Threats of shortages after the oil crisis of 1973. 153

8.3 An engineer working on an analogue calculator at the research and development center in Neder-over-Heembeek, 1969. 159
8.4 Claude Loutrel, Jacques Solvay, and Whitson Sadler at Deer Park plant, Texas. 161
8.5 Karol Wojtyla – Pope John Paul II – visits Rosignano plant in 1982. 168
9.1 Solar Impulse. 170
9.2 Solvay recovers Bernburg plant in 1991. 175
9.3 Ten-leva Bulgarian banknotes featuring Devnya soda ash plant acquired by Solvay in 1996. 177
9.4 CEO Daniel Janssen meeting King Bhumibol of Thailand, together with Belgian Ambassador Patrick Nothomb (1998). 179
9.5 The negotiators of the Solvay-BP deal in 2001. 185
9.6 Jean-Pierre Clamadieu, successor of Christian Jourquin as the head of Solvay (2012). 189
9.7 Nicolas Boël, Chairman of the Board of Solvay since 2012. 190

FOREWORD

Solvay is a company in which Science is highly respected. Therefore I suggested to Solvay S.A. that professional historians should write a book at the time of the company's 150th anniversary.

The present volume, as well as the scholarly book *Solvay: History of a Multinational Family Firm*, published at the same time,[1] are the result of an exciting research project that took place over five years. The challenge was taken up by a remarkable team of international historians (Kenneth Bertrams, Nicolas Coupain, and Ernst Homburg, acting under the outstanding and discreet leadership of Ginette Kurgan-van Hentenryk). These historians benefited from the insights provided by an "Industrial Committee" (Aloïs Michielsen, Jean-Marie Solvay, Jacques Lévy-Morelle, and me), as well as by many present and past Solvay managers. We thank them wholeheartedly.

Besides our respect for historical science, we were interested to show the determining influence of History on the life of Solvay between 1863 and 2013.

May Solvay employees and shareholders, as well as any interested readers worldwide, enjoy this lively and scientific historical book.

Daniel Janssen
Honorary Chairman Solvay S.A.

---

[1] Kenneth Bertrams, Nicolas Coupain, and Ernst Homburg. *Solvay: History of a Multinational Family Firm*. Cambridge: Cambridge University Press, 2013.

Introduction

A Company in History: Solvay, 1863–2013 deals with the history of the Solvay company from a broad perspective. It intends to show how, and to what extent, the history of a multinational family company was rooted in and a product of 150 years of world history.

If students were asked to write an assignment on "Solvay" in 2012, they would immediately start their research on Wikipedia, the most popular online encyclopedia of the time. Redirected to the "disambiguation page," which lists different articles with the same title, the students would have to choose between seven headings:

- Solvay (company): an international chemicals and plastics company
- the Solvay process
- Ernest Solvay, its inventor
- Solvay Conference
- The Solvay Business School
- Solvay, New York
- Solvay Hut, on the Matterhorn in the Alps

Yet this list would be valid for English-speaking researchers only. In German, several headings would be missing, but two new items would appear: "Solvay GmbH," as the German chemical company part of the Solvay group and, surprisingly, "Solvay (7537)" – an asteroid discovered by the Belgian astronomer Eric W. Elst at the La Silla Observatory in Chile in 1996. Italian (and Slovene) students would also be aware of the asteroid, but they could also learn that the town "Rosignano Solvay" owes its name to the soda plant set up by Solvay & Cie in the vicinity of Rosignano Marittimo in 1913 (although contributors failed to mention that the name "Solvay" was added by local authorities in the 1920s). A well-illustrated article explains the history and style of the factory's model village – "Villaggio Solvay" – described as a fine and rare example of a garden city in Italy. Finally, French-speaking students would not be able to find out about the Solvay refuge in the Alps, nor about the asteroid. Instead, their artistic curiosity might expand from learning about the Hôtel Solvay designed by architect Victor Horta and a hallmark of the Art Nouveau style on the Avenue Louise of Brussels. A further heading could also lead them to an article on the Solvay Library at the Parc Léopold in Brussels, originally built as Ernest Solvay's Institute of Sociology in 1902.

One name thus refers to many people, places, and things. Nevertheless, before being a village, an asteroid, or even a chain of mountains of the Antarctic Peninsula, a setting that Wikipedia failed to mention, Solvay was the name of a family, a family whose legacy derives from one of its outstanding members – Ernest Solvay. Following his early experiments in manufacturing soda with ammonia, the chemical company he founded in 1863 with his brother Alfred and with the help of several partners became a remarkable achievement in the world chemical industry. The reader interested in understanding the unfolding of the company's successive "lives," from its origins through the celebration of its 150th anniversary, will find thorough information in another book published at the same time as the current volume.[1] As already noted, this book's ambition consists in recasting the company's history in broader terms. Of course, this utopian goal would be unrealistic unless the author made some major choices in what to include; that is, much had to be left out. Therefore, special emphasis is placed on adopting a general comparative stance as a means to single out Solvay's peculiarities but also its common tendencies in a global environment. Last but not least, it should be stressed that these pages, which address

---

[1] Kenneth Bertrams, Nicolas Coupain, and Ernst Homburg, *Solvay: History of a Multinational Family Firm*, New York: Cambridge University Press, 2013. Hereinafter, *Solvay*.

many issues already familiar to scholars and students of history, have been written for a general readership curious about what is currently associated with the word *Solvay*, be it a world-famous scientific conference, a multinational company, or that company's founder. To some extent, this essay will have achieved its task if it succeeds in bringing out the common meaning of all things bearing the name Solvay, assuming a part of the answer lies in their convergence in modern history. And this also applies to the asteroid #7537.

This book owes its very existence to the research carried out by my colleagues Nicolas Coupain and Ernst Homburg. They not only have improved previous versions of the manuscript, but they also allowed me to literally plunder their findings. Words fail to express my gratitude and friendship. I am sincerely grateful to my estimate colleague Ginette Kurgan-van Hentenryk, as well as to the members of the Industrial Committee – Daniel Janssen, Aloïs Michielsen, Jean-Marie Solvay, and Jacques Lévy-Morelle – for their tireless efforts in making this book accessible to a wider audience. Speaking of form improvement, I thank Shana Meyer and her team for their expert polishing of my English in the final draft of the book. I am particularly indebted to Nicolas Coupain for the selection of figures and the writing of captions. Finally, my last and special thanks go to Flavia and Mathias – *al tempo che abbiamo perso e a quello che recupereremo*. This book is dedicated to the past and present workers of the Solvay company.

K. B.

Figure 1.1. Machinery in Motion court, International Exhibition, London, 1862. Millions of visitors could discover there the latest industrial progress coming from participating Nations. (Science Museum/Science & Society Picture Library.)

I

The Golden Age of Progress

– My dear Sir, that's the whole question. There is the only difficulty that science need now seek to overcome. The problem is not how to guide the balloon, but how to take it up and down without expending the gas which is its strength, its life-blood, its soul, if I may use the expression.
– You are right, my dear doctor; but this problem is not yet solved; this means has not yet been discovered.
– I beg your pardon, it has been discovered.
– By whom?
– By me!
– By you?

Jules Verne
Five Weeks in a Balloon, 1863

OPENING: DOCTOR FERGUSON, MISTER SOLVAY

Jules Verne's novel *Five Weeks in a Balloon* tells the story of an ingenious English scholar, Dr. Samuel Ferguson, intending to travel across the still mysterious continent of Africa in a hot air balloon. A device of his invention, a mechanism of five receptacles allowing for the combustion of hydrogen gas at different temperatures, enables him to stay in the air for a long time without the need to release gas or drop ballast to control the balloon's altitude. Traveling westward from Zanzibar to Senegal, Ferguson and his two companions fly over unfamiliar regions of Africa and face many dangers. After an epic odyssey, they ultimately manage to return to England where they establish, "in the most precise manner, the facts and geographical surveys" reported by previous explorers.

The account, filled with suspense, exoticism, and a dash of technology, is typical of the adventure novels that would bring international fame to their author. More than (science) fiction, however, Verne's story is an invaluable testimony of his time and of his contemporaries – at least a small fraction of them. In 1863, the world was in expansion; Verne's readers admired the industrial development, technological improvements, and scientific advances of the day, as well as the conquest of civilization over "uncivilized" peoples. The lay prophets of this religion called progress were scientists, inventors, and explorers – no wonder the main character of Verne's novel combined all these highly esteemed vocations. Members of society's upper strata would rush to public lectures to learn of these innovators' research and findings, presented at prestigious scholarly societies (such as the famous Royal Geographic Society, founded in 1830). Examples of the achievements of these visionaries were revealed at international exhibitions, which attracted tens of thousands of spectators from London (in 1851 and 1862) to Paris (1855 and 1867) up to the huge Philadelphia Centennial of 1876. Despite their differences, these scientists, inventors, and explorers shared the belief that existing knowledge could be challenged and, it was hoped, improved (or even proven wrong). In this sense, they performed an act of rebellion. "In every town, nay almost every village, there are learned persons running to and fro with electrical machines, galvanic through-holes, retorts, crucibles, and geologist hammers," observed an Englishman as early as 1828.[1] For many candidates, however – perhaps for the bulk of them – the attempt to challenge what existed would eventually lead to failure and renunciation. Success was scarce and therefore extremely appealing to creative spirits.

Ernest Solvay was one of these creative spirits. He certainly belonged to the category of "enthusiasts, [who were] realists and dreamers at the same time" (to use his own words).[2] Had he not been so restrictive in the use of his spare time, he might have been

---

[1] Ian Inkster, *Science and Technology in History: An Approach to Industrial Development*, New Brunswick, NJ, Rutgers University Press, 1991, 287.
[2] Ernest Solvay, "Industrie et science (Biogénie et sociologie)," *Revue scientifique*, XLVIII (2nd semestre), 1910, 705–11 (at p. 705).

reading *Five Weeks in a Balloon* when it came out in 1863. However, for some time, Ernest Solvay had decided to devote the rare hours of his time off at his uncle Florimond Semet's factory to undertake chemical experiments. Supervising production at a gas works was not the most pleasant activity available on the job market (nor was it the worst, to be frank), but it had unsuspected advantages for an amateur chemist: It was a place where a product like ammonia was easily obtainable, even wasted as by-product, in coal distillation. As a result, Semet's factory, located in the vicinity of Brussels, became Ernest Solvay's research laboratory as much as it was his playground – the setting of his experiments as a gifted and inspired tinkerer.

The rest of the story could unfold as a traditional, fairy tale–like success story: Solvay would have discovered the ammonia-soda process, set up the thriving company to exploit and commercialize the product that resulted from it, and become a wealthy tycoon reigning at the top of his industrial empire. History, however, is far from being a continuous flow of successes, let alone fairy tales. Solvay's is a true story full of failures, nuance, and blind spots. What is striking in the early stages of Solvay's enterprise is that the story evolves like Jules Verne's account of the balloon flying above Africa: upward with cheerful enthusiasm, then downward nearing total collapse, and then up into the sky again.

The company Ernest Solvay and his partners built (for Ernest was not alone in this endeavor) was finally established after many failed attempts. For years, Solvay & Cie, where *Cie* stands for Compagnie (Company), was a small-scale start-up on the brink of bankruptcy. Capital was lacking, industrial output was dragging, and business partners were nagging. More important, Solvay did not discover the soda-ammonia process; he *rediscovered* it yet thought for long that he was the first to make it happen. Besides, although he eventually became rich, Ernest Solvay was much more than a mere businessman; he devoted his time and energy to countless initiatives, many of which lay beyond the sphere of industry.

SETTING THE STAGE: THE TRIUMPH OF INDUSTRIAL CAPITALISM

What kind of world, what kind of society, unfolded before Ernest Solvay's eyes as he started his professional life at his uncle's factory in the early 1860s? It was, to be sure, a world in the midst of profound change. At the eve of the nineteenth century, only a few observers contested the assumption of endless progress. The English scholar Thomas Malthus was one of them. In his treatise *An Essay of the Principle of Population* (published in various editions between 1798 and 1826), he argued that the geometric growth of population would soon outstrip the mere arithmetic growth of resources. If radical changes were not taken, famines, diseases, and decline were looming. Yet Malthus's concern proved wrong. By the middle of the nineteenth century, European regions were swept away by an unprecedented wave of economic expansion. Malthus's own country, Great Britain, was in the pole position of this process. The mid-Victorian period saw Britain ruling the world not only in terms of imperial leadership but also as the uncontested "workshop of the world." In coal, iron, and steel production, it accounted for almost half of the total world output between the 1840s and the 1860s, which were Britain's "golden years."[3] The iconic symbol of this flagrant domination was the development of the railway industry. Here again, Britain was the figurehead of a movement that would spread worldwide through massive export of railway iron and machinery. In a few decades, Europe and North America became two webs of steel. The railway mileage on each one of these continents grew from approximately 2,000 in 1840 to 30,000 in 1860, and more than 100,000 in 1880. By that time, however, eighteen European countries were equipped with railway lines; thirty years earlier, they were only nine.[4] This was a remarkable achievement, felt so even by those who lived it. If Jules Verne's readers trusted that the world was in expansion, they also believed this gigantic transformation in communication and transportation enabled world unification. As remote parts of the globe increasingly became linked to rest of the world, the scope of the economy shifted from international to global.

Quite naturally, the process of industrialization affected the workforce, working patterns, and society as a whole. Widening of the labor market followed the expansion of the market for consumer

[3] Martin Daunton, *Wealth and Welfare. An Economic and Social History of Britain, 1851–1951*, Oxford, Oxford University Press, 168–70.
[4] Eric Hobsbawm, *The Age of Capital, 1848–1875*, London, Weidenfeld & Nicolson, 1975, 54–5.

goods, creating a society of "mass production." At the same time, urbanization was reaching new levels; capitals and other major cities were attracting both the petite bourgeoisie and a cohort of unskilled workers. Of course, this took time. Traditional economic and societal forms – agriculture, consumer shops, craft industry, and the like – were still overwhelmingly present. During this time, Alexandre Solvay, Ernest's father, started working as a teacher but shifted toward the more lucrative occupations in the trade and quarry business. The size of the enterprise was modest but sufficient for him to move his way up through the milieus of the local bourgeoisie. To some extent, the same business format applied to Florimond Semet's gas works in Brussels. Small and medium-sized firms were the norm; large was the exception.[5] Gradually, however, larger units of production took over, especially in the "heavy" industrial sectors. When it prevailed, the factory system did not mushroom everywhere. It was concentrated in local areas rather than nations as a whole. These industrial districts – the Midlands and Lancashire in Britain, the Borinage in Belgium, the Ruhr in Germany – consisted of industrial hives surrounded by overcrowded dwellings and slums. They were the hidden and hideous backstage of economic prosperity. As one historian put it, "The world was therefore divided into a smaller part in which 'progress' was indigenous and another much larger part in which it came as a foreign conqueror, assisted by minorities of local collaborators."[6]

Belgium was a prime mover in terms of industrialization. In fact, by the middle of the nineteenth century, it ranked second in the world, just after Great Britain. The country accounted for a number of blast furnaces and stationary steam engines at a proportion that found no equivalent among its continental neighbors. Means of transportation, railways and waterways in particular, were well developed. Since its independence in 1830, the small nation was committed to the ideology of liberalism and industrial "progress." Under the impulsion of King Leopold the First and his son, Leopold II, it was a hotbed for private initiatives and the home of a stateless "self-help" society (the development of the socialist

[5] Arno Mayer, *The Persistence of the Old Regime*, New York, Pantheon Books, 1981, 37–44.
[6] Eric Hobsbawm, *The Age of Empire, 1875–1914*, London, Weidenfeld & Nicolson, 1987, 31.

movement was only truly influential after 1880). Like Britain, Belgium was firmly anchored to the principles of free trade and international exchange. With France, Italy, and Switzerland, it adhered to the Latin Monetary Union in 1865, a kind of prototype for the "euro zone" monetary system. A good illustration of Belgium's commitment to the mechanisms of liberal economy was its flexible patent legislation enacted in 1854. The Belgian patent system provided a fairly long period of validity (twenty years) compared with other countries but did not require any form of preliminary "paternity check."[7] This administrative carelessness gave way to unexpected outcomes, as Solvay's case highlights.

THE BURNING SUN OF KNOWLEDGE AND THE ART OF NOT KNOWING

If the industrial economy was booming – even with the outbreak of violent depressions through the century – it was owed, at least in part, to knowledge and innovation. The philosopher Alfred Whitehead once observed that "the greatest invention of the nineteenth century was the invention of the method of invention."[8] The judgment was certainly more valid for the last third of that century than for the preceding years. What happened from the 1870s onward that was so important was a closer interaction between science and technology for industrial purposes; more precisely, technology became more science-based, and industry tended to rely increasingly on technology-oriented knowledge or applied science. True, before that time, technical inventions did not involve much thorough scientific knowledge. The industry rested on pragmatic knowledge and know-how, trial-and-error experiments, and empirical craftsmanship. Slowly but surely, however, things started to change. After 1850, "science became more important as a handmaiden of technology."[9] Steel-industry magnate Andrew Carnegie marveled at the wonder of trained chemists for the quality

---

[7] Gabriel Galvez-Behar, *La République des inventeurs. Propriété et organisation de l'innovation en France (1791–1922)*, Rennes, Presses Universitaires de Rennes, 2008, 34.
[8] Alfred N. Whitehead, *Science and the Modern World*, Cambridge, Cambridge University Press, 1926, 11.
[9] Joel Mokyr, *The Lever of Riches. Technological Creativity and Economic Progress*, Oxford, Oxford University Press, 1990, 113.

improvement of metallurgical products: "Nine-tenths of all the uncertainties of pig iron making were dispelled under the burning sun of chemical knowledge," he later claimed.[10] Although this was more a change of degree than in nature, the period has been called the Second Industrial Revolution. Pivotal, then, were the rise and diffusion of a new set of inventions related, for example, to steel, electricity, and (organic) chemicals, as well as new power sources, especially automotive engines based on internal combustion. The point here is that these elements were not only innovative technologies; they have been the starting point of a new dynamics in industrial development.

Ernest Solvay's early experiments occurred during the prehistory of the Second Industrial Revolution. By many accounts, his research practices still pertained to the serendipitous and "learning-by-doing" style of invention characteristic of the first half of the nineteenth century. He was both a child of his time and an actor in the change to come, and this he had in common with many others inventors. An interesting example is the English businessman Henry Bessemer, who in 1856 introduced a groundbreaking technique for the manufacture of steel – using a "converter" as a means to blow air directly through the molten metal. Bessemer was "a kind of high-class tinkerer already wealthy by his ingenuity and versatility."[11] Like Solvay, he had no formal training, but his invention gave way to the rise of metallurgy (or materials science) as a new discipline for engineers at the borders between science and technology.[12] The prehistory of the Second Industrial Revolution, which Solvay and Bessemer embody, was thus remarkably intense in terms of technological and scientific innovations. It has even been claimed that "if one had to choose any fifteen-year period in history on the basis of the density of scientific breakthroughs that took place, it would be difficult to find one that exceeded 1859–1873" (1859 corresponds to the publication of Charles Darwin's *Origins of Species*

---

[10] Quoted in David Mowery and Nathan Rosenberg, *Technology and the Pursuit of Economic Growth*, Cambridge, Cambridge University Press, 1989, 30.
[11] David Landes, *The Prometheus Unbound. Technical Change and Industrial Development in Western Europe from 1750 to the Present*, Cambridge, Cambridge University Press, 1969, 255.
[12] Donald Caldwell, *The Norton History of Technology*, New York, Norton, 1995, 291–2.

Figure 1.2. Louis-Philippe Acheroy, childhood friend of the Solvay brothers and first employee of the firm at the Schaerbeek test plant. He is shown here in his workshop at Couillet, where he was head of production. (Solvay Archives)

and 1873 to James Clark Maxwell's textbook *A Treatise on Electricity and Magnetism*, which provided the theoretical grounds to electricity-based industries).[13]

---

[13] Mowery and Rosenberg, *Technology and the Pursuit*, 22–3.

Knowledge and innovation were thus highly praised values at the time of Ernest Solvay's discovery, but when he filed his first patent titled "Industrial Production of Soda Ash with Sea Salt, Ammonia and Carbonic Acid" on 15 April 1861 (one day before his twenty-third birthday), Solvay had no clue that a dozen chemists had found "his" invention before him. It is only when the first semi-industrial steps of his experimental station had yielded promising results and he was in search of business partners that Solvay became aware of this anteriority after consulting the patents collections of the Musée de l'Industrie in Brussels. "It was a hard blow for me," he is supposed to have said – the kind of blow that prompted him to quit.[14] Solvay's biographers all point to this event – the discovery of nondiscovery – as a turning point in the future of the industrial process and its success. Just as there is a "Eureka moment" at the core of many inventors' narratives,[15] the epic and drama of Solvay's success story is built on a reverse angle – ignorance (and how to overcome it). Not knowing, in other words, had been an important drive for Solvay until his "violent disappointment" of 1862 (to use the words of his sister-in-law). Instrumental, then, was the support of Solvay's legal adviser and partner-to-be, Eudore Pirmez, who apparently persuaded him to move forward. A new strategy ensued from this disillusionment: patenting everything, everywhere. Solvay and Pirmez began to file a patent for each stage of technical operation and each new device required for manufacture at the industrial scale. They did so preemptively in France and England as well. The fundamentals of the process were not patented because the chemical reaction, as Solvay admitted belatedly in his second (1863) patent, had been "known for a long time."[16]

MERRY INVENTORS, RELUCTANT ENTREPRENEURS

For the majority of successful inventors, patenting a new device or mechanism was simply a necessary administrative step to attain

---

[14] [René Purnal] *Vie d'Ernest Solvay*, Bruxelles, Lamertin, 1929, 29; Marie Solvay-Masson (wife of Alfred Solvay), *Les débuts de la Société Solvay & Cie, mémoire intime*, Internal Solvay publication, Brussels, May 1915, 23.
[15] Thomas P. Hughes, *American Genesis. A Century of Invention and Technological Enthusiasm, 1870–1970*, Chicago, University of Chicago Press, 1989, 75.
[16] Quoted in Bertrams, Coupain, and Homburg, *Solvay*, 21, n. 44.

the intellectual property of their invention. Bureaucratic constraints notwithstanding, it was not a difficult task. Exploiting the new product full scale, on the other hand, required a wide array of skills and contacts. It was perceived as a real challenge, a journey into the unknown. Hence, many inventors preferred to opt out of that part and sell or license the rights of their patents to entrepreneurs who manufactured and marketed the inventions at their own risk. Here again, however, there were exceptions – the so-called inventor-entrepreneurs. Famous names in this group include Thomas Edison, Alfred Nobel, Alexander Graham Bell, and Giuseppe Marconi. Today they are mostly remembered as belonging to the pantheon of genius inventors; the collective memory has largely overlooked, if not completely erased, their entrepreneurial achievements. It is true that the organizational action of these independent inventor-entrepreneurs has been overshadowed by the system that took over their initiatives at the turn of the twentieth century – namely the development and dominance of corporate research and development (R&D) laboratories.

Contrary to widespread intuition, however, successful inventors who launched their own business ventures rarely did so with the same youthful eagerness that had prevailed during the process of invention. Even the most celebrated of them all, Thomas Edison, who systematically designed the companies that exploited his inventions, left the task of merging with Thomson-Houston Electric Company (derived from another inventor-entrepreneur, Elihu Thomson) to the managers acting under the leadership of the awesome financier J. Pierpont Morgan. General Electric Company, which resulted from this sensational merger in 1892, became a giant concern in the new electric-light systems industry, its leadership disputed only by its eternal rival, Westinghouse Electric Company. Other examples of this initial entrepreneurial reluctance abound (consider also the creation of Bell Telephone Company in 1877). As one historian put it, independent inventor-entrepreneurs "reserved their enthusiasm and primary creative thrust for the act of invention; they performed the entrepreneurial function of establishing companies because they wanted to bring their invention into use."[17] Inventions, in other

---

[17] Hughes, *American Genesis*, 24.

Figure 1.3. Chemical works at Floreffe (Belgium) in about 1850. Ernest Solvay considered selling his process to this company before launching his own business. It is also in this plant that the lawyer Eudore Pirmez familiarized himself with the chemical industry before entering into contact with Ernest Solvay. (Lithograph from Bart Van Der Herten, Michel Oris, and Jan Roegiers, eds. *La Belgique industrielle en 1850: Deux cent images d'un monde nouveau*. Antwerp: MIM/Crédit Communal, 1995. Original lithograph by Jules Géruzet, 1850–1855; printed by Simonau & Toovey; artist: Adrien Canelle)

words, came first, commercialization second, and only by way of necessity.

Surprising as they may sound, these stories square broadly with Ernest Solvay's early entrepreneurial steps. Thanks to new evidence,[18] we now know that in 1862 he was on the verge of selling the rights of his patent to a Belgian chemical and glassmaking company. When the deal fell apart because the initial patent was invalid, industrial partners were sought (e.g., the French company Saint-Gobain), but these efforts did not pay off. Against all odds,

---

[18] Bertrams, Coupain, and Homburg, *Solvay*, 20.

Solvay's entourage opted for the creation of a limited-partnership company funded with capital raised among partners – family members and owners of local small-scale businesses. Solvay & Cie was officially launched on 26 December 1863 with a modest starting budget (leading to constant money worries), but financial pitfalls were just one of its many shortcomings. Shortly after the issues of intellectual property and corporate legal structures were overcome, technological obstacles cropped up. These were more serious because they put the very nature of the process and its industrial outcomes in jeopardy. From its inception in 1864, the small plant built in Couillet near Charleroi, located in the heart of Belgium's glassmaking industrial district, was the theater of a succession of technical mishaps. Thus, the factory running the soda-ammonia process did not yield the expected quantity of soda ash. "It is not possible that we [will] always have bad luck," wrote younger brother Alfred to Ernest Solvay. Alfred had given up his commercial training in Hull, England, to help him develop the production at Couillet, together with their childhood friend, Louis Acheroy.[19]

Alfred Solvay was right: misfortunes began to fade progressively after 1867. Development, growth, and expansion came next. Strikingly, the Solvays never gave up their firm belief in the industrial exploitation of the ammonia-soda process. A hunch became conviction, and then obstinacy. In retrospect, the episode of successive failures at Couillet became Ernest and Alfred Solvay's own road to Damascus, something they never forgot and of which they were keen to remind others (it would also be extensively covered in their later hagiography as a means to legitimate their success). To an English industrialist willing to run his process, Ernest Solvay wrote in 1876:

May I draw your serious attention upon the importance of details in our manufacturing process? You may need up to six months to understand it... but you will easily realize it step by step. Details, the constant care [of technicalities] have more importance to us than great matters. I advise you to progressively proceed to laboratory and factory testing. Do every

---

[19] Eliane Gubin and Valérie Piette, "Une histoire de familles," in Andrée Despy-Meyer and Didier Devriese (eds.), *Ernest Solvay et son temps*, Brussels, Ed. Archives de l'ULB, 1997, 95–136 (at p. 113).

day something else; otherwise you may lose it all. Tough times [have] only [begun] for you.[20]

This excerpt highlights another important fact. As in most chemical and electrical companies of the time – and in the overwhelming majority of the industrial economy for that matter – the procedures conducted at Couillet were truly and fundamentally unspectacular. They consisted in the day-to-day work of production, routine testing, and quality control. The nature of the working practices of the Second Industrial Revolution was adaptive rather than revolutionary.[21] Adjusting the temperature of ovens and coolers, cleaning the filters of sodium bicarbonate, and emptying the distillation vessels are scarcely exciting tasks, but they were of utmost importance for the manufacturing process and the eventual success of the enterprise. Despite their thankless nature, they formed the core of what could be called a neglected form of industrial innovation.

SELF-MADE MEN DO WORK IN A COLLECTIVE ENVIRONMENT

The Solvay brothers' obstinacy, however, lay not only in the meticulous supervision of their equipment. Like many of his counterparts, Ernest Solvay persevered in his enterprise despite reticent, if not negative, judgment about the validity of his process from some of his relatives as well as top-notch scientists of his time. Much like Edison and other fellow inventor-entrepreneurs (Bell was an exception), Solvay could not provide expert credentials to support what he was doing and where he was going – hence, the need, usually spurred on by the inventor's business partners, to resort to some form of scientific authority. It was a kind of a competition between two forms of knowledge – trial-and-error experimentation on the one hand and abstract theory on the other. Roughly speaking, it also involved two groups of people – self-made tinkerers versus educated scholars. If Carnegie, as we have seen, admired the input

[20] Cheshire Record Office, Archives of Brunner, Mond Company, DIC/BM 19/24, Ernest Solvay to James Richards, 16 May 1876.
[21] Robert Fox and Anna Guagnini. *Laboratories, Workshops, and Sites. Concepts and Practices of Research in Industrial Europe, 1800–1914*, Berkeley, Office for History of Science and Technology, University of California at Berkeley, 1999, 173–4.

of "the burning sun of chemical knowledge" for his industrial goals, scientific knowledge could not always be at the forefront of technology. Furthermore, academic expertise, although far from being the essence of stereotyped "ivory towers," was seldom the best guide for creative minds. Successful inventor-entrepreneurs are at the vanguard of scientific exploration. Science, in contrast, is the information available at a given time, and scientists are mostly concerned with proven science. "I can answer questions very easily after they are asked," a university graduate said to Thomas Edison, "but find great trouble in framing any to answer."[22]

At the early stages of his experiments, Ernest Solvay was confronted with the dire predictions of acknowledged and respected professors. "Men of pure science are generally too rigorous in their judgment," he later said.[23] Not only did Solvay persist in his efforts despite their negative warnings, he also continued to rely on university-based science and seek out academic consultants once he and his staff had scaled up their industrial production. As his later life would amply show, he never became resentful about academics and universities, scientists and science, and this episode confirmed the apparently paradoxical saying that self-made men *do* work in a collective environment, whether socially and practically.[24] If one were to take a picture of the creative "human factors" in action during the process of invention leading to commercialization, it would undoubtedly be a group portrait that would bring together relatives, friends, technicians, partners, consultants, financiers, politicians, and a whole cohort of off-screen go-betweens. The density of human traffic engaged at the various stages of the working process demonstrates that we are miles away from the cliché of lonely geniuses isolated from the rest of the world and mocked by their contemporaries.

---

[22] Hughes, *American Genesis*, 47.
[23] Ernest Solvay, "Coup d'œil rétrospectif sur le procédé de fabrication de soude à l'ammoniaque," reprinted under the title "L'histoire d'une invention," *Revue de l'Université Libre de Bruxelles*, VIII, 1902–1903, 591–602 (at p. 599).
[24] W. Bernard Carlson, "Innovation and the Modern Corporation. From Heroic Invention to Industrial Science," in J. Krige and D. Pestre (eds.), *Companion to Science in the Twentieth Century*, London, Routledge, 1997, 203–26 (at p. 207).

The Golden Age of Progress 19

Figure 1.4. Thomas Edison in his lab in Menlo Park, New Jersey. Edison embodied the inventor-entrepreneur of the nineteenth century. Ernest Solvay possessed the same characteristics. (Mary Evans Picture Library / Classic Stock / C. P. Cushing)

Although a solitary character, Ernest Solvay was not alone. Evolving in a dense family web, he benefited from his parents' and relatives' direct and intermediate personal connections that, in the long run, contributed to no less than four kinds of "capital": financial, social, scientific, and technological. The scope and ramifications of Solvay's social dynamics are addressed in the next chapter.

Suffice it to say now that such personal networks were largely co-constituent to the success of the enterprise, but some of the closest among them were also present in times of crisis. The meaning of the expression "for better or for worse" would soon become clear to Adèle Winderickx. She was only eighteen years old when Ernest Solvay married her on 17 September 1863. The notion of a family business applied for Solvay & Cie because the company and family histories were closely intertwined, and the overlap between private and professional spheres was not concealed. Among other examples, the commemoration of the company's fiftieth anniversary was celebrated on the very day of Adèle and Ernest's golden wedding anniversary in 1913.[25]

There was more to this extension of social capital, however: in addition to people, a wide array of equipment was also involved in the collective environment. Another popular belief depicts inventor-entrepreneurs as if they must, per force, exercise their creative drive in rundown environments. This was hardly the case for most: the bulk of inventor-entrepreneurs worked in settings specifically designed for the purpose of problem solving. Of course, the Couillet plant was not an academic laboratory; it was just a shack compared with Thomas Edison's invention complex, established in 1876 in Menlo Park, New Jersey. Nonetheless, despite the rather scant technology available, the factory contained the necessary logistics for the conduct of manufacturing and "unspectacular" innovation. With a dash of drama, Alfred Solvay's wife, Marie, described in a nutshell the collective atmosphere of the Solvay brothers' working conditions:

The training of the personnel as a whole, the manufacture of new devices, the practical and economic exploitation of a process that was so simple in theory but so delicate to operate, everything had to be undertaken by young people who were neither engineers nor chemists and were struggling with limited financial resources.[26]

Ernest Solvay fully admitted his lack of higher technical education: "I am neither a chemist, nor an engineer," he once told qualified

---

[25] Gubin and Piette, "Une histoire de familles," 136.
[26] Solvay-Masson, *Les débuts de la Société Solvay & Cie*, 32.

chemists at a conference in Berlin in 1903.[27] Except for formal training, however, he could not deny that his efforts, whether successful or not, were embedded in a much larger collective framework. The truth is that most of the members of this network have been either neglected or perceived as ancillary instruments to his objectives.

SODA ASH AND THE GHOST OF NICOLAS LEBLANC

An industrial colossus lay at the heart of the Second Industrial Revolution – the science-based industry. Simultaneous to the groundbreaking innovations of the inventor-entrepreneurs, a new set of discoveries took place directly in in-house industrial laboratories or involved joint industry-university partnerships. Electricity and (organic) chemistry were the prime sectors affected by this transformation, which also saw a transfer of geographic nexus. Originated in England, the industry soon moved to Germany, which became the seedbed of science-based innovation, together with the United States, in the last quarter of the nineteenth century. It is precisely in the field of high added-value chemicals that German companies distinguished themselves and would soon thrive. Bayer and Hoechst, both created in 1863 (like Solvay & Cie), as well as BASF, established in 1865, expanded on their ability to "capture" outside knowledge and develop strategic planning of research-and-development to harness markets. Americans were not lagging behind, however. The new breed of research-intensive firms included Du Pont and Kodak in the chemical sector, as well as General Electric, Westinghouse, and American Telephone & Telegraph Company in the electric-supply and telecommunications industries. The corporate laboratories these companies set up employed highly qualified researchers, some of whom would be awarded a Nobel Prize. The emergence of the science-based industry has undeniably profoundly changed the dynamics of knowledge, technology, and industrial capitalism.[28]

---

[27] Ernest Solvay, "Coup d'œil rétrospectif," p. 593.
[28] The literature on this issue is countless, from Georg Meyer-Thurow's classical "The Industrialization of Invention. A Case Study in Chemical Industry," *Isis*, 73, 1982, 363–81, to François Caron's recent synthesis *La dynamique de l'innovation. Changement technique et changement social (XVIe–XXe siècles)*, Paris, Gallimard, 2010, esp. 220–30.

Figure 1.5. Semi-industrial laboratory at Saint-Gilles (Brussels), around 1900. From the 1880s on, the research carried out in this lab aimed to improve existing processes as well as investigate new areas of development, such as electrolysis. (Solvay Archives)

Given what we have already discussed, it would be inaccurate to list the ammonia-soda industry in the category of the science-based industry, nor were the main German synthetic-dye manufacturers contenders of Solvay & Cie in the market of chemicals. They were not playing in the same league. Obviously, the challenge for Solvay lay in the existing alkali industry running the Leblanc process. Unlike Ernest Solvay, who eventually became a prosperous businessman, the French surgeon Nicolas Leblanc was an ingenious inventor but an unsuccessful entrepreneur. Ruined, he committed suicide in 1806. Despite the human tendency to reject and forget failures (his name is not engraved on the Eiffel Tower among the seventy-two prominent French scientists and technicians of the eighteenth and nineteenth centuries), his contribution deserves more than a footnote in the textbooks of history. For his process of manufacturing soda from common sea salt, Leblanc was issued one of the first patents of the French Republic in 1791. More important, his invention opened the era of substitute chemicals, which were of utmost importance for the development of nineteenth-century industry, especially for overseas exports.[29] Ironically, the process became a major source of industrial activity sixty years after the death of its inventor. The rewards of Leblanc's efforts were reaped by those who exploited his idea. "The twenty years between 1860 and 1880," a historian wrote, "were the golden age of the Leblanc industry."[30] By the mid-1860s, the manufactures running the Leblanc process employed some 10,000 men in England and Wales. The Leblanc industry was stimulated by the soaring demand for textiles, soap, and new varieties of paper, which brought about increased use of bleaching powder.[31]

This formed the background from which, and against which, Ernest Solvay and his staff started their venture – conquering the alkali market led by the Leblanc manufacturers. The risk was immense and failure the most probable outcome. No wonder Ernest Solvay was haunted by the ghost of Nicolas Leblanc (he would later have a sculpture of Leblanc in his office). But the Solvay team had come to terms with the idea and reality of failure after technological obstacles had been overcome. Their faith in the industrial success of the soda-ammonia process was inexorable. The process was simpler and more elegant than that of its rival. Also, at a time when environmental issues were just beginning to emerge (the Alkali Act was passed in Britain in 1863 as a means to lower the polluted air produced by Leblanc works), the Solvay process represented a promising opportunity. Above all, however, it was cheaper and led to continuous production. With these ingredients, Solvay & Cie took its first steps into the international alkali market.

---

[29] Bernadette Bensaude-Vincent and Isabelle Stengers, *Histoire de la chimie*, Paris, La Découverte, 1992, 208.
[30] L. F. Haber, *The Chemical Industry during the Nineteenth-Century*, Oxford, Clarendon Press, 1958, 95.
[31] Landes, *The Prometheus Unbound*, 271.

Figure 2.1. Four generations of Solvay leaders: Ernest (1838–1922), Armand (1865–1930), Ernest-John (1895–1972), and Jacques (1920–2010). (Solvay Archives)

2

Outline of a Family Business

> You are a Buddenbrook. We are not born to live for what we might take to be happiness with our short-sighted eyes. We are links in a chain. You, too, are unthinkable without those who have come before us.
>
> Thomas Mann
> *Buddenbrooks*, 1901

CRITERIA AND COMMONPLACES

Question: What do Coca-Cola, Hyundai, and IBM have in common?

Answer: They are all family-controlled companies.

Unlike Solvay, Ford, or Hewlett-Packard, not all family-controlled corporations have the family name tagged on their companies or their products. The Woodruff family still wields huge influence in the Coca-Cola Company, the Watsons at IBM, and the Ju-yung dynasty holds the reins at the Hyundai Group.

Nothing is more common than a family firm. The family structure has been the backbone of the history of economy ever since the invention of capitalism. From the sea merchants of the ancient world to the crafts-based artisans of urban gilds, from the labor-intensive workshops of the Industrial Age to the current networks of the postindustrial society, the family has been and still is an obvious model for the setting of an enterprise. Yet the more one looks into details in the specific characteristics of a family company, the more one is puzzled by the lack of homogeneity of this very notion. Defining the concept of a family business is not an easy task. Is it a matter of ownership, or a matter of control? And if it is a matter of control – especially relevant in the case of listed companies – on which field should that control be essentially exerted: capital or production, day-by-day decision making, or general strategy? Does it have to do with the size or the nature of the branch? Or is it after all a question of style, the so-called corporate culture, that defines a family business? Does it, then, depend on the organizational patterns, or on the long-term chain of succession? Indeed, there are as many criteria as there are exceptions to what makes a firm a family firm. One would better opt for a smooth and subjective understanding of the notion.

Aside from this inexorable search for the one and only defining feature, the family business is often associated with a series of impressions and prejudices, more or less based on observations, that contribute to shape its understanding. For instance, common sense would claim that family companies are first and foremost small-scale companies specialized in, say, shop keeping, luxury goods, or craftsmanship. They are supposed to be absent when it comes to drafting a list of major corporations. Another perception, partly perpetuated by historians and economists, argues that the "family business" is a rather old-fashioned, if not outdated, structural model that petered out during the early stages of the industrialized society. In other words, the family unit has not resisted the rise of managerial capitalism with the increasing role played by outside professionals at the command of enterprises. A final assertion is that family businesses are not successful, and hence, they are not long-lasting. Despite the ongoing rhetoric about dynasty and durability, which concerns a handful of such businesses, family firms supposedly often fail the test of the second or third generation. The latter is especially evident in the well-known "Buddenbrooks effect," which stems from Thomas Mann's literary portrait of a well-off nineteenth-century merchant family. Because of a crisis in the succession from the second to the third generation, the Buddenbrooks family began its slow but inexorable decline as the company was engulfed in an abyssal collapse.

Such cheerful and optimistic perspectives on family businesses have long prevailed. They are either exaggerated or plainly false.

Table 2.1. World's largest family-owned firms in 2004 (by Revenues)

Firm	Country	Industry	Percent owned or controlled by family
Wal-Mart	United States	Retailing	Walton family owns approximately 38%
Ford Motor	United States	Automotive	Family holds 40% of voting shares
Samsung Group	South Korea	Electronics	Lee family controls approximately 22%
LG Group	South Korea	Electronics	Koo and Huh families own approximately 59%
Carrefour Group	France	Retailing	Families control 29% of voting shares
Cargill	United States	Agriculture/food	Family owns approximately 85%
Fiat Group	Italy	Automotive	Agnelli family owns approximately 30%
PSA Peugeot Citroën	France	Automotive	Peugeot family holds 42% of voting shares
BMW	Germany	Automotive	Quandt family controls 47% of shares

Source: *The Economist*, 4 November 2004.

Family companies have not been doomed to be small-scale, old-fashioned, or less profitable than their nonfamilial competitors. In fact, the model of personal capitalism enjoys a great impulse, which is perceptible in early-industrialized countries as well as in today's emerging markets. Whatever the business sector, currently some of the world's largest firms are either family-owned or family-controlled (see Table 2.1). When it comes to big business, estimates suggest that 17 percent of the top 100 American and German corporations at the end of the twentieth century were family companies.[1] Other calculations indicate a much higher proportion in France (26%) and Italy (43%).[2] On the other hand, statistics hardly reflect the reality at work when one knows how discreet, yet decisive, the influence of families can be in the

---

[1] Andrea Colli, *The History of Family Business, 1850–2000*, Cambridge, Cambridge University Press, 2003, 16.
[2] Harold James, *Family Capitalism*, Cambridge, MA, Belknap Press, 2006, 4–5.

boardrooms of some of the largest firms. Furthermore, the frequent use of multiple voting rights in big companies makes it difficult to distinguish what really lies behind the vague notion of "influence."

Solvay has been, and still is, a vibrant member of the family of family businesses. Through its 150-year history, the company has succeeded in maintaining the mark of the Solvay family in terms of both control but and short- and long-term decision making. This family stamp does not imply, however, that the company was bound to be utterly conservative and dogmatically wary of non-family stakeholders or outside managers, as we will see. Nor did the succession process focus on selecting family scions irrespective of their competence. Nevertheless, it remains an undisputable truth that the "family touch" of the Solvay company has been an essential ingredient of its success, as well as a crucial component of its stability and survival. The next pages cast light on how the singular rediscovery of a young inventor-entrepreneur actually formed the core of what became a family business called Solvay & Cie.

THE FIRST AND CLOSEST CIRCLE – THE SOLVAY CLAN

Like fairy tales, histories of families usually start with a "once-upon-a-time" kind of bait intending to arouse the imagination of the reader.[3] In the case of the Solvays, it would be even more relevant to combine the saying with a "once upon a place," for genealogy and geography are closely intertwined in understanding the company's early development. In the eighteenth century, the Solvays settled in the village of Rebecq, located twenty miles south of Brussels. The town did not lie in the heart of the booming Belgian industrial center but belonged to a rural region, and its semi-industrial takeoff started only in the mid-nineteenth century. To a large extent, however, Rebecq is for the Solvays what the village of Hayange in Lorraine is for the well-known Wendel family – a dynastic home, the site where it all began. Rebecq is not only Ernest Solvay and his siblings'

---

[3] This section draws extensively on Eliane Gubin and Valérie Piette, "Une histoire de familles," in Andrée Despy-Meyer and Didier Devriese (eds.), *Ernest Solvay et son temps*, Brussels, Ed. Archives de l'ULB, 1997, 95–136.

Figure 2.2. Four Solvay presidents: standing: Aloïs Michielsen (CEO 1998–2006; Chairman of the Board 2006–2012), Daniel Janssen (CEO 1986–1998; Chairman of the Board 1998–2006); seated: Yves Boël (Chairman of the Board 1991–1998), Jacques Solvay (CEO 1971–1986; Chairman of the Board 1971–1991). Among them, Aloïs Michielsen is the sole manager – and the first CEO – who was not from the Solvay family. (Solvay Archives)

birthplace; it is also where the Solvay company took shape in social terms because of the family's various connections. For the sake of clarification, three circles can be delineated in terms of degree of kinship and intimacy: family, relatives, and friendships.

Figure 2.3. Porphyry quarries of Quenast. Porphyry represented one of the major industrial activities of the rural region of Rebecq. Alexandre Solvay, father of Ernest and Alfred, managed one of these quarries in the 1840s and 1850s. (Lithograph from Bart Van Der Herten, Michel Oris, and Jan Roegiers, eds., *La Belgique industrielle en 1850: Deux cent images d'un monde nouveau*. Antwerp, Belgium: MIM/Crédit Communal, 1995. Original lithograph edited by Jules Géruzet in 1850–1855; printed by Simonau & Toovey; artist: Adrien Canelle)

The first circle, of course, concerns the closest family members, which could be labeled the "Solvay clan." Alexandre Solvay (1799–1889), Ernest's father, immediately comes to mind as the family patriarch – a term he would perhaps not have discarded. The youngest of a family of eight children, Alexandre swiftly resigned from the teaching position he and his brothers had inherited at the Rebecq boarding room created by his father in 1793. Obviously, this was not the kind of living he had dreamed of. He was lured by the promising development of trade and business, a field in which some of his acquaintances had started their careers. He opted for the nearby quarry industry and obtained a concession for exploitation there. The quarry was called Renaissance, a relevant name considering Alexandre's fresh ambitions, and the move was undoubtedly a smart one. Porphyry-based ballasts were required for the stabilizing of railroads and, since their introduction in Belgium in 1835, railways enjoyed an enduring expansion. But 1835 was an important year in another respect, as Alexandre married Adèle Hulin (1809–1878) that year. The eldest daughter of a family of five, Adèle came from a relatively wealthy family – wealthier family than the Solvays, at any rate, especially if one takes into account the nature and size of real estate. In a sense, Alexandre and Adèle's wedding paved the way to a closer interweaving of the families, and Adèle's two younger brothers married two of Alexandre's sisters in the 1850s. With future generations carrying on the wedding tradition, the genealogy of the Solvays and Hulins became so intricate that years later, in 1915, Alfred Solvay's wife, Marie, called it a real "puzzle for current descendants."[4] The moral legacy that Alexandre and Adèle Solvay cultivated for their five children was one typical of a midrange bourgeois society, subscribing to the liberal values of education, progress, and work, tinged with religious beliefs yet wary of clerical dogma. Alexandre's intellectual mind set was encapsulated by a series of "thoughts and maxims" that were later gathered and made available to the family by his eldest son. Some of these aphorisms seem relevant in view of what was to come: "The happiest man is the busiest man," "Success stems from perseverance," and "The greatest men's glory should always be measured through the means they have used to obtain it."[5]

If one focuses on the young Ernest Solvay while considering the first circle, two persons should be highlighted at the core of this inner circle: Ernest's younger brother, Alfred (1840–1894), and his wife, Adèle Solvay born Winderickx (1845–1928). Their role in Ernest's life, as well as in the creation and subsequent development of the family business, is proportionally inverse to the credit and acknowledgment they have received for it. Alfred was Ernest's first

---

[4] Marie Solvay-Masson (wife of Alfred Solvay), *Les débuts de la Société Solvay & Cie, mémoire intime*, Internal Solvay publication, Brussels, May 1915, 11.
[5] *Pensées et Maximes glanées par Alexandre Solvay*, Brussels, privately printed, 1900, 58, 62, 95.

partner and best friend. Of course, the teamwork of brothers is far from exceptional in the history of industrial capitalism. What seems striking in the case of the two Solvays is that the younger brother was not a late-coming adjunct partner (as was Joseph Thyssen for his "big brother" August) or a professional dedicated to a function that was assigned to him (as was Anton Philips for a few years before he became partner). Although Ernest gave the general directions, it was Alfred who, together with Louis Acheroy (discussed subsequently), did much of the behind-the-scenes experimental work at the rudimentary testing station in Brussels, as well as at the Couillet plant when Solvay & Cie was set up. In this sense, Alfred was, with Louis Acheroy, the company's first-ever salaried employee; he earned a monthly wage of 2,500 Belgian Francs. Years later, when Solvay & Cie was thriving, Alfred's involvement in the day-to-day business did not wane – quite the contrary. Personnel throughout the company praised his human and organizational skills. When he died prematurely in 1894, Ernest became more irritable and isolated than ever, as his biographers unanimously agree. He was determined to evoke Alfred's role in all circumstances. Later on, successive generations of managing partners at Solvay & Cie frequently referred to Ernest and Alfred as "the founding brothers" (*les fondateurs*) as a means to perpetuate the legacy of collective entrepreneurship. Today's collective memory, however, has notably and unduly diminished Alfred in the overall picture, and he is not the only one on whom this fate has fallen.

If the Solvay family story were a movie, Adèle Solvay-Winderickx could perhaps have received an academy award in the category of best supporting actress. Her story is a real one, however, but, not unlike most women in business history – and in history more generally – it is has been considered only in her role as Madame Solvay. Then again, just as for Alfred, the life and deeds of Ernest's wife have been outshined by that of her husband. This is perhaps because her role was strictly confined by, and entrenched in, the moral values shaped by late-nineteenth-century bourgeois society. Before becoming Madame Solvay in 1863, Adèle Winderickx was already part of Ernest's family: she and Ernest shared the same grandparents on the side of their respective mothers – Jean-Joseph Hulin and Lucie Marie Cooreman. Hence, Adèle knew well the Solvay's Rebecq network. The family dowry and legacy she brought to the marriage was an undeniable source of capital for her husband, especially welcome in the early years of struggle. Besides, Adèle's father, Anselme Winderickx, never failed to provide his son-in-law "moral and material relief" when he badly needed it for his industrial venture.[6] One should not overlook the fact that, precisely during the so-called difficult years of the business, Adèle was raising the couple's four children: Jeanne (1864–1947), Armand (1865–1930), Hélène (1866–1938), and Edmond (1870–1940). At first sight, the distribution of roles gives the impression of a rather gender-based division of labor – Ernest at the plant, Adèle at home. This might be accurate and in line with the requisites of the bourgeois environment of their time. Nonetheless, Adèle was working on both sides – starting a family and, at the same time, contributing to the development of an industrial dynasty. Often in Ernest Solvay's professional and personal preoccupations, she was in charge. Among her many initiatives, her involvement in the successive real estate acquisitions of the Solvay clan, in and around Brussels, has frequently been pointed out. Her dual roles could rightly serve as a metaphor to characterize her temperament as a whole: a supporting hand with her feet firmly rooted to the ground.

THE SECOND AND THIRD CIRCLES: RELATIVES AND FRIENDS

Through his union with Adèle Hulin, Alexandre Solvay not only guaranteed an upgrade in lifestyle; it also extended his and his children's social capital into a second circle of Solvay associates comprising indirect relatives, most of whom were in-laws. Among them were the Semets, whose industrial experience and success were well established. As already noted, Florimond Semet, Adèle's brother-in-law, owned the gas works that later became Ernest Solvay's first experimental site. A generation later, Louis Semet, presumably Florimond's nephew, married Ernest's youngest sister, Elisa, in 1874 and became managing partner at Solvay & Cie for some twenty years. An engineer with great experimental skills, he devised a system of coke ovens that supplied ammonia as by-product; the

---

[6] Solvay-Masson, *Les débuts de la Société Solvay & Cie*, 32.

technique was evidently instrumental for the continuous production trend of the ammonia-soda process. In fact, many of Ernest's in-laws proved resourceful, including the physician Léopold Querton, who married Aurélie Solvay, and the notary Alphonse Delwart, the husband of Alphonsa Solvay, whose family was already historically linked to the Solvays. At one stage or another, their offspring would occupy an important position in the family company. On the other hand, belonging to the circle of relatives – as to the other circles for that matter – was not necessarily a guarantee of success. Take, for instance, the changing disposition of Florimond Semet toward his nephew when the latter started to neglect his job at the gas works. It eventually led to the (forced) resignation of Ernest Solvay from his uncle's company in August 1865, and with that the guarantee of a position with good wages was gone. Ernest had to devote himself full-time to his brainchild, whatever the risk for him and his family. At the time, nobody could predict that this was a blessing in disguise for the future of Solvay & Cie. Likewise, thirty years later, an American chief engineer born in 1863 (definitely a good year for entrepreneurship) followed the same hazardous pathway as he quit his job at Edison Illuminating in Detroit. He wanted to sell automobiles. His name was Henry Ford.

The second circle was not composed exclusively of in-laws. It also included a close friend, already mentioned several times, who, despite his not being a formal family member ought nonetheless to be placed in that category. To historians, Louis-Philippe Acheroy remains an enigmatic and mysterious figure. He is a mere shadow to the other players in the Solvay story. The member of a family native to Rebecq, he gave up his initial vocation in the holy orders to follow the more risky paths of the chemical industry suggested by his childhood friend, Ernest. He first became the assistant to Alfred at the testing station in Brussels and pursued the technical direction of production once the company was running full steam. Without any formal technical education, he "graduated" summa cum laude from the learning-by-doing school of experience, spending most of his professional life on the shop floor. Over time, Acheroy became an essential part of the company. His technical expertise in the ammonia-soda process, from its general design to the technicalities, made him a highly qualified worker in terms of industrial drawings and process control. For that reason, he was frequently asked to ensure the supervisory control of foreign and new-build soda plants. Until his departure from the company in 1894, his loyalty to the Solvay brothers was immense. There is no doubt that a significant part of the company's success is owed to this discreet, hardworking friend.

The third and final circle comprised early friends and supporters of the Solvays without whom the family business would never have seen the light of day. The first in the list was a longtime genuine friend of the Solvays, Guillaume Nélis (1802–1896), a member of the local industrial notability. The owner of a regional paper mill factory and stakeholder of several chemical companies, he was also a member of the Chamber of Deputies. For this reason, he represented an invaluable doorway toward upper-class society and its potential for seed money. It was Nélis who introduced Ernest Solvay to a fellow representative at the chamber – Eudore Pirmez (1830–1890). Depicted as a brilliant corporate lawyer and an elegant orator, which would eventually lead him to a seat in the Belgian government (1868–1870), Pirmez belonged to one of the most prestigious industrial dynasties of the Charleroi region.[7] As mentioned in Chapter 1, he made Ernest Solvay aware of the need to obtain irrefutable paternity of his invention. And once proof of Ernest's priority was disputed, Pirmez found the means to exhort him not to give up. While Ernest Solvay thought in terms of technological breakthroughs, Pirmez's philosophy was that of an investor: if so many had risked so much on that invention, it meant that a huge profit was at stake. "It thrills me more than it frightens me," is what he is said to have declared at the time.[8] As a result, Pirmez took the lead of the fundraising campaign. He first asked his family members to step in. He convinced his father (Léonard Pirmez), his aunt (Hyacinthe Pirmez), and his father-in-law (Valentin

---

[7] Jean-Louis Delaet and Valérie Montens, "Famille Pirmez," in Ginette Kurgan-van Hentenryk, Serge Jaumain, and Valérie Montens (eds.), *Dictionnaire des patrons en Belgique*, Bruxelles, De Boeck, 1996, 513–17.

[8] Quoted in Jean-Louis van Belle, *Solvay & Cie. Recueil d'archives inédites relatives à la société en commandite, 1862–1890*, Braine-le-Château, Ed. La Taille d'Aulme, 2008, 36. See also Jacques Bolle, *Solvay, L'invention, l'homme, l'entreprise industrielle*, Bruxelles: Ed. Weissenbruch, 1963, 50.

Figure 2.4. Solvay managers visited the production sites regularly to stay in close contact with the workforce. From left to right: Édouard Hannon, Hönnecken, Armand Solvay, Nicolas de Krukoff, Fernand van der Straeten, Ernest Solvay, Alfred Solvay, and Léopold Querton visiting the Wyhlen soda ash plant in Germany, around 1890. (Solvay Archives)

Lambert) to invest in the start-up. He then reached out to another of his and Guillaume Nélis's colleagues in Parliament, Gustave Sabatier (1819–1894), who was also manager of coal-mining plants. Interestingly, Nélis, Pirmez, and Sabatier were all members of the Belgian Liberal Party, then labeled the "industrial party" because it staunchly promoted a laissez-faire economic policy.[9] But they joined the deed to the word. The project that Ernest Solvay so convincingly brought to them was exactly what they were looking for to further industrial expansion.

---

[9] Ginette Kurgan-van Hentenryk, "Gustave Sabatier," in Kurgan-van Hentenryk et al., *Dictionnaire des patrons*, 539–40.

LIFE IN A PARTNERSHIP COMPANY

The transition from the rediscovery of the soda-ammonia process to the enterprise itself was not linear, nor was it self-evident. Although Solvay & Cie could be described from the outset as a family business in terms of firsthand development and managerial culture, the notion did not apply when it came to ownership: no less than three-quarters of the initial capital were put up by partners outside the family. This was the norm rather than the exception at the time. Entrepreneurs were desperately seeking associates to fuel their creations with funding. The alternative would have been knocking at the doors of investment banks – and it would inevitably have yielded even worse loan conditions. Like all entrepreneurs, Ernest Solvay had no illusion when it came to financiers. As a high-ranking executive at the Peugeot & Frères automobile company once said, "A banker is someone who reluctantly lends you a parasol when the sun shines, but refuses an umbrella as soon as it starts to rain."[10] Therefore, the story of Solvay & Cie as seen from the perspective of the early business partnership is an interesting one. As an epic drama, the chain of events unfolds in three successive acts.[11]

Not surprisingly, the launching of Solvay & Cie in December 1863 constitutes the first act. "At this stage precisely, my story shifts in nature," wrote Alfred Solvay's widow in her memoir on the company's origins, "it looks like a business relationship rather than a family narrative."[12] Indeed, the scale of investment – and, thus, of ownership – was at the core of the discussions surrounding the acquisition of capital for the venture and involved the Solvays and the components of the third circle described earlier. Whereas the Solvay brothers received thirty-four shares representing the "approximate cost" of the Brussels testing station and its equipment, as well as the estimated value of the patents and the process, the other associates – Nélis, Sabatier, and the Pirmez family – held the bulk of the shares. And the shares they owned were

---

[10] Quoted in David Landes, *Dynasties. Fortunes and Misfortunes of the World's Great Family Businesses*, London, Viking, 2006, 175.
[11] Unless noted, this section draws on Bertrams, Coupain, and Homburg, *Solvay*, chaps. 1 and 3.
[12] Solvay-Masson, *Les débuts de la Société Solvay & Cie*, 24.

listed in a first-class category of privileged stock, which meant that they had priority in the distribution of earnings. Last but not least, Solvay & Cie was legally incorporated as a stock-based limited partnership, the most common corporate enactment of the time. It implied a clear distinction between managing partners (*gérants commandités*) on the one hand, and silent partners on the other. While the *gérants* were responsible of the management and liable for all the debts of the business (up to their personal fortune), the silent partners were not involved in the management and were liable only for debts incurred by the firm to the extent of their registered investment. In sum, the Solvay brothers bore the risk of the project without being rewarded for it. However unfair the system seemed, it was the price to pay for obtaining venture capital. As previously noted, the only viable alternative at the time were banking institutions providing high-interest-rate loans.

The second act of the limited partnership coincided with the quasi death of the business (Scene I) immediately followed by its shaky revival (Scene II). As expected, the first years of Solvay & Cie put the partnership to a harsh test of resistance. With an unlikely, even nonexistent, prospect of profitability, the firm was on the verge of collapse. Silent partners, conversely, intended not to stay silent anymore. The pressure reached a climax when, at a shareholders meeting held in Rebecq on 29 October 1865, Alexandre Solvay heard about the company's level of indebtedness. When he further learned that, in a limited partnership, bankruptcy had unlimited consequences for the managing partners, he reluctantly agreed to borrow more money as a means to stabilize the finance. After lengthy negotiations, it was agreed to give Solvay & Cie a final chance. A new arrangement was drafted in 1866 that reflected the changing conditions of the partnership. Although the Solvay family now held a stunning share of the capital, holding no less than 20 percent of the stock, the bylaws of the company were even more restrictive on the Solvay brothers. As majority shareholders, in contrast, the Pirmez and Lambert families could have conformed to the expectation of "absentee ownership" of the time. They did not, however, and decided to pitch in. Silent partners suggested technical and commercial improvements; they were also looking for expert advices among established chemists, and were in constant search for further investors.[13] By and large, this collective, innovative environment allowed the company to move out of the abyss. More symbolically, the revival of the business overlapped, in the late 1860s, with the moving of Ernest and Adèle Solvay (and their three children) to the Rue du Prince Royal in Ixelles, in the southeast of Brussels. Their new home was the beginning of their real estate expansion on the same street, where later the central headquarters of the Solvay industrial empire would reside. Much later, in 1883, the family would move in a magnificent property located Rue des Champs-Elysées, which would become the Maison Ernest Solvay.[14]

The third act of the business partnership was in many ways the most surprising and unpredictable. Contrary to those that preceded it, this act took place at a much later stage, namely throughout the 1880s, when Solvay & Cie was a thriving international business. In the meantime, technical legal adaptations had been implemented. More important, however, the Solvay family started off its striking ascent in the company's ownership and control. This was done by overcoming a series of technical obstacles (capital increases, creative accounting, etc.), which had required an expertise in the art of corporate legal engineering. What appeared to be a strategy of *reconquista*, a kind of a take-back program, culminated with the revision of the company's bylaws in 1882. The Solvays now held some 64 percent of the partnership shares (Ernest alone held more than the half of it), with which an unequal proportion of voting rights was associated. For Ernest, this spectacular turnaround was only natural. He and his brother had "worked themselves to death" for the company's success, as he later justified. Accordingly, he trusted the *gérants* deserved some of the profit they had been deprived of in the earlier versions of the bylaws. The silent partners, needless to say, opposed this view. A lawsuit ensued from the deadlock of negotiations between managing and silent partners. The litigation ultimately ended when a financial settlement was reached in 1890. Eudore Pirmez, who had taken the lead of the opposing party, did not outlive these transactions; he died before age sixty after an intense professional life. His last salient activities in the

---

[13] Van Belle, *Solvay & Cie*, 36, 167.
[14] Gubin and Piette, "Une histoire de familles," 120.

political arena had consisted in chairing the Higher Council of King Leopold's Independent State of Congo after the 1885 Berlin Congress and the Belgian Labor Committee, set up after the industrial riots around Liege in 1886. Pirmez's death concluded the business partnership's three-act drama. The company's structure was left untouched until it went public in 1967. Silent partners remained important shareholders and were always represented in the company's supervisory board, which controlled its financial health.

OUTSIDERS NEEDED (OR HOW TO AVOID KING LEAR SYNDROME)

When in 1888 the time came to celebrate the twenty-fifth anniversary of Solvay & Cie, there was no doubt that Ernest Solvay had not only become a captain of industry, he was also the sole captain of his vessel. Nevertheless, by the mid-1880s, his fragile health, mainly the result of what we would today call work-related burnout that would later turn into chronic nervous breakdowns, forced him to relinquish the reins of his "masterpiece." The idea of transforming the limited partnership into a joint-stock company, while preserving the family's upperhand through various shareholding mechanisms, had been raised on many occasions. The advantages were many, but the fear of being taken over by a third party had always been of primary concern to Ernest Solvay. He was not opposed, however, to diversification of the company's management. As Shakespeare's *King Lear* tragically demonstrated, heredity alone did not suffice to maximize performance. It is not certain whether Ernest Solvay ever expressed a taste for the classics of English literature, but he surely knew how to conduct a business. Early in the company's history, he made clear his intention to inject new blood into the venture. At Solvay & Cie, the first prototype of extra-Solvay management was named Prosper Hanrez (1842–1920). The son of a successful industrialist specialized in the manufacturing of engines, Hanrez was a civil engineer trained at the respected Mining School of the University of Liege.[15] To a large extent, his profile radically differed from that of the Solvay brothers – and that is why they

[15] René Brion, "Famille Hanrez," in Kurgan et al., *Dictionnaire des patrons*, 349–51.

hired him in 1873. He became the designer, organizer, and plant manager of Solvay & Cie's first-ever soda plant located outside Belgium, at Dombasle in the French Lorraine. The plant outshone everyone's expectations, and Dombasle became a model for other Solvay factories to come. As a reward for his achievements, Hanrez was appointed managing partner in 1879 and received a percentage of the company's stockholdings. But he wanted more – more than the brothers were willing to give. The clash with Ernest became inevitable, and Hanrez resigned from Solvay & Cie in 1886.

Hanrez was replaced by Louis Semet, Ernest and Alfred's brother-in-law. If anything, the Hanrez experience had proven that loyalty to the company must prevail over individual ambitions. And in this respect, nothing could provide more certainty than being part of the family, even if a certain "leeway" was required to ascertain the competence of the candidates. If there was a real gem who possessed both the loyalty of a family member and the competence of a professional manager, then it might as well be found among the armada of sons-in-law. As a result, the second circle's successive generations were mobilized to integrate the highest echelon of the company's managerial hierarchy – the managing board, or *gérance*. In-laws had the unquestionable advantage of combining family and nonfamily assets, being insiders and outsiders at the same time. The entrance of new *gérants* occurred in two successive waves. The first occurred in 1894 to deal with the sudden loss of Alfred at age fifty-four. Ernest's eldest son, Armand, was called to the board. He was joined there by Fernand van der Straeten (1856–1932), the husband of Ernest's daughter, Hélène. The second wave took place in 1907, an important year for the revitalization of Solvay & Cie's management as it saw the integration of no less than four new *gérants*. Three of these displayed a whole range of family ties: Ernest's cousin (Edgard Hulin, 1851–1919), the son of Ernest's brother-in-law (Georges Querton, 1874–1914), and Alfred's only son (Louis Solvay, 1876–1952). The fourth managing partner and only nonfamily member was the engineer Edouard Hannon (1853–1931), who had been hired at Solvay & Cie back in 1876 and had next been appointed head of the company's technical department.

Interestingly, the habit of co-opting in-laws was carried out intensively after the outbreak of the First World War, and no

Figure 2.5. Dinner at Carl Wessel's (Deutsche Solvay Werke manager), around 1900. Seated on the left side, notably: the family managers Georges Querton, Edgard Hulin, and Fernand van der Straeten surrounding Carl Wessel; on the right side, Ernest Solvay and Louis Semet. (Solvay Archives)

outsider was admitted to the managing board until Solvay & Cie became public as Solvay S.A. in 1967. After that year, the appointment of non–family members either on the Board of Directors or the Executive Committee was more common, albeit not a frequent practice. This is indicative of the importance of stabilizing the company around the backbone of the family, conceived *lato sensu*. As we have seen, Ernest Solvay had carefully avoided the threat of capital dilution. By the same token, the family has managed to overcome the well-known pitfalls of family business succession to the second and third generations, or "Buddenbrooks effect." Following Ernest's death in 1922, the potential outbreak of a "war of families" in terms of corporate control and managerial structure was averted, in part due to a renewed focus on family ties and family values. Mutatis mutandis, the same pattern had already been applied by other important family businesses, such as in the Wendel group. After the death of the patriarch, Charles de Wendel, in 1784, the question of the company's unity and survival was explicitly raised, and it was eventually decided to pursue its future in the following terms: "With M. de Wendel, the House was a family in a company, and without M. de Wendel the House must be a company in a family."[16] Successful successions, in other words, are necessary for the sustainability of family businesses. Necessary as they are, however, these conditions do not suffice to explain what makes a family business a thriving multinational company, and to this topic we now turn.

---

[16] Quoted in James, *Family Capitalism*, 139.

Figure 3.1. Pumping engine at the Solvayhall plant, Germany. In this country, Deutsche Solvay Werke exploited potash deposits in the Stassfurt area. (Solvay Archives)

3

Building an Industrial Empire

> More and more, interests are diversifying, concentrated, interwoven; beyond borders and races, beyond barriers of trade, there operate the giant concerns of industrial and financial capitalism.... While banks are funding distant subsidiaries in every land beyond the seas, they also rise, concentrate capital and interlock interests, in such a way that a stitch of credit once torn in Paris provokes a sudden collapse in Hamburg and New York. Hence, it gives birth to a capitalist solidarity, which can be dreadful when it is orchestrated by inferior spirits, but which can become the guarantee for peace under the inspiration of the common will of peoples.
>
> Jean Jaurès
> Speech to the Parliament, 20 December 1911[1]

"FASTER, HIGHER, STRONGER": PICTURING THE FIRST GLOBALIZATION

The last quarter of the nineteenth century, up until the outbreak of the First World War in 1914, saw an extraordinary expansion of the world economy, with Europe at its core. The remarkable feature of this period of growth was that it did not rest, first and foremost, on a quantitative foundation. Certainly, the volume of trade, the degree of industrial output, and the quantities of energy supply, overwhelmingly based on coal and steam power, were rising

---

[1] Quoted in Suzanne Berger, *Notre première mondialisation*, Paris, Seuil, 2003, 83.

sharply. Recent estimates point to a growth of 492 percent in real trade in Belgium between 1870 and 1914 compared with a weighted European average of 294 percent.[2] However, the striking aspect of the changes lay elsewhere – namely, in the intensity of the process, the extension of the environment where it unfolded, and the diversity with which it materialized. "Faster, higher, stronger" could have been the slogan of this accelerated version of capitalism. In fact, this was the Olympic motto – "Citius, Altius, Fortius" – introduced by Pierre de Coubertin when he set up the International Olympic Committee in 1894. Not surprisingly, the word "better" was absent from this slogan. If the living standards of the working classes were generally improving compared with the dreadful situation of the early nineteenth century, the fruits of prosperity were still inadequately distributed – the more so when one considers the distribution of wealth from an international perspective: the world, then, was divided between winners and losers. In some regions, such as the industrial valley of Liege in 1886, this social unbalance led to an unprecedented outburst of protests and riots. Elsewhere in the industrialized world, organized unions and political parties started to effectively represent and defend the interests of the workers striving for an improvement of working conditions.

The political context is another striking analogy to the reintroduction of the Olympics, but it is more of a reverse analogy. Whereas the modern Olympic ideal rested on the principle of internationalism, feelings of nationalism were simultaneously at the forefront of the political landscape in Europe and beyond. Of course, nationalist movements were far from novel; the various ideological impulses for the formation of independent nation-states date back to the revolutionary era of the eighteenth century. Nonetheless, the nationalist momentum that rose from the last quarter of the nineteenth century onward was slightly different from previous undertakings. Potential nation-builders now predominantly based their argument on ethnical and language-based criteria. According to the new breed

---

[2] Guillaume Daudin, Matthias Morys, and Kevin H. O'Rourke, "Globalization, 1870–1914," in S. Broadberry and K. Hh. O'Rourke (eds.), *The Cambridge Economy History of Modern Europe, Vol. 2: 1870 to the Present*, Cambridge, Cambridge University Press, 2010, 5–29 (at p. 7).

of nationalist advocates, common ethnic-linguistic values enabled "nationalities" to become independent "nations."[3] A minority of these common values might have actually existed, but the bulk of them were the result of creative imagination in the wake of the long tradition of romanticized literature.[4] Whatever the sources of the "national question" actually were, breaking free from the old empires to which these nationalities still belonged became a sport much en vogue in late-nineteenth-century Europe. Despite its ever-increasing popularity, however, the discipline of nationalism was never granted Olympic recognition.

As a new firm in quest of stabilization, Solvay & Cie had to deal with this rapidly changing world. Given the nature of the company's flagship product, soda ash, the political and economic context was not something vague and remote, something that was part of general considerations. For many reasons, international (geo)politics and economic relations played a role in the company's early strategy. An obvious explanation was that the manufacturing of soda ash according to the Solvay process required raw materials (salt, lime, ammonia, and coke) that were to be found in, or provided by, various locations. Among these basic products, salt was by far the most important. The Couillet plant, which was not located near a salt deposit, could thus not become the company's model or reference factory. But there was more. International expansion was justified as an essential condition for Solvay & Cie's growth, and even its long-term survival. A mass consumer product rather than a niche merchandise, soda ash had to be sold in large markets. Besides, the potential the Solvay process offered, especially in reducing the production costs, represented a window of opportunity that could be better met on an international level, even in a highly competitive market. Overall, from the rationalization of manufacturing capacities to the evidence of worldwide market opportunities, an array of reasons prompted Solvay & Cie to face what could be called the necessity of international expansion.

---

[3] Eric Hobsbawm, *Nations and Nationalism since 1780. Programme, Myth, Reality*, Cambridge, Cambridge University Press, 1990, 101–103.
[4] Benedict Anderson, *Imagined Communities. Reflections on the Origin and Spread of Nationalism*, London, Verso, 1983, 83–111.

Figure 3.2. Bags of sodium carbonate (soda ash) at the Italian plant of Rosignano. (Solvay Archives)

However, ensuring the success of this expansion strategy was by no means an easy task. The company had to take up three challenges. First, Solvay & Cie was an industrial dwarf compared with the predominance of Leblanc manufacturers, particularly in Great Britain. This represented the burden of being a challenger, not the reigning leader. The favorable aspect of this situation, however, was that Solvay & Cie could catch industrial leaders by surprise, which is what it did to the Leblanc producers. Second, the company had to come to terms with the limitations of its own financial resources. Family firms certainly have many advantages, but the access to instant capital is usually not one of them. At the same time, the industrial landscape witnessed the swift emergence of large managerial enterprises backed by the stock market or investment banks, especially in the United States, and their contrast with the family business model was striking. This represented the challenge of scale. The third dimension was perhaps more subtle but no less pervasive. It concerned the unexpected outburst of a great depression from the early 1870s onward that was due mainly to industrial overproduction. This depression affected international trade as a whole and paved the way to a long period of protective tariffs (until and even after the First World War) that hindered export-oriented industries.[5] As it happened, this unfavorable "economic trend" coincided with Solvay & Cie's first steps in the big world. It surely did not prevent the company from surfing on the waves of globalization, inasmuch as Solvay & Cie was manufacturing a cheaper product compared with its competitors. However, it contributed to undermining the international flow of trade. This was the challenge of timing.

We now know that Solvay & Cie overcame the obstacles presented by these three challenges and managed to steer its way through international expansion while maintaining its initial family structure. What is less well known is that the successive steps taken toward internationalization were not linear. Further, this driving force for expansion led to a variety of results in terms of foreign development. Solvay & Cie did not stretch its arms with equal speed and strength regardless of the local constraints at stake in France, England, Germany, or elsewhere. This is to say that the image we might have of a multinational company that unfolded like a gigantic web around the world is largely misleading. Of course, Solvay & Cie's path to expansion started from Brussels, but, early on, it occurred to Ernest and Alfred Solvay that their business had to deal with, and adjust to, national cultures of capitalism. Often this meant engaging in a battle that took place on the frontlines of business and politics, which were strongly intertwined.

On the business side, the strategy seemingly consisted of taming competitors rather than squeezing them. Ernest Solvay was

[5] Hobsbawm, *Age of Empire*, 38–9.

straightforward on this matter: "Intending to force Leblanc manufacturers to reduce too fast their production capacities amounts to stimulating them to set up themselves soda plants running the soda-ammonia process."[6] Negotiation, according to the founder, was thus the core strategy chosen for Solvay & Cie's industrial policy. Nevertheless, this was a long way from being "competitor-friendly." It was achieved through significant and systematic price reductions, which would be pursued to ensure the slow decline of the Leblanc process. Political issues, on the other hand, were far from negligible. The Solvays quickly realized the harshness of national (i.e., nationalist) reactions to foreign-based entrepreneurship. Needless to say, they were not the only industrialists affected by the political constraints of the time. Analyzing the development of Thomas Edison's electricity system overseas in the 1880s, the historian Thomas Hughes observed that the results of the technology transfer through a strategy of patent licensing were strikingly divergent from country to country. He noted that "the most penetrating explanation for the failure in London and the success in Berlin is neither technological nor economic; it is political."[7] At first sight, the struggle came down to a reenactment of the political clash between empires and nation-states in the business arena. This was true to a great extent. For, much like political empires, industrial empires could transform themselves into quasi-national components.

HOW TO BECOME FRENCH (IN FRANCE) AND GERMAN (IN GERMANY)

Solvay's relationship with France had started early in its history, with the importation of salt from the Lorraine region to supply the Couillet plant. In fact, for a long time, the rationale of Solvay's "French connection" would revolve around Lorraine and salt, salt in Lorraine. When by 1870 Ernest Solvay and his associates started to search for a proper location to establish a factory in this region of France, international politics came into the picture: the

---

[6] Archives Centrales Solvay (ACS), "La politique commerciale des frères Solvay," 28 July 1938, 2.
[7] Thomas P. Hughes, *Networks of Power. Electrification in Western Society, 1880–1930*, Baltimore, Johns Hopkins University Press, 1983, 77.

Franco-Prussian War broke out. Soon the provinces of Alsace and Lorraine were occupied by Prussian troops, which victoriously reached Paris and marched even beyond the capital. The armistice was signed on 28 January 1871. It was a disaster for the Second French Empire, which lost Alsace and part of Lorraine and became the Third French Republic after a quasi civil war. It was a triumph for the Kingdom of Prussia, which subsequently gave birth to the German Empire under the impulsion of its new chancellor, Otto von Bismarck – one empire down, another one up. Still, this major geopolitical turmoil gave way to minor local outcomes as far as Solvay & Cie was concerned: the "salty" part of Lorraine the Solvays were interested in remained French territory after the war was over. Local authorities were keen on facilitating the arrival of investors. The Belgian company was allowed to buy a piece of land in the town of Dombasle in October 1872; the soda plant was erected in the following years with a first unit on stream in December 1874. Dombasle factory comprised seven units in 1882 producing up to 210,000 tons soda ash per year (a figure achieved in 1912). Impressively, the plant's performance was also visible on the country's balance of trade. From 1883 onward, France became a net exporter of soda ash, and the presence of Solvay & Cie on its soil had made the difference. This success was largely due to the relentless efforts of the plant manager, Prosper Hanrez, whose role in the Solvay group was inestimable.

As one would expect, Solvay & Cie's stunning entry in the French soda market did not remain uncontested.[8] Leblanc manufacturers were eager to curb the intruder's enthusiasm by any means necessary. French industrialists knew too well that Solvay & Cie relied heavily on the local production of salt. In 1877, Lorraine salt producers grouped and exerted a strong pressure toward price increase. Solvay & Cie replied by buying out its own salt concessions, enabling the direct supply of its essential raw material. In turn, salt producers like Marchéville-Daguin, Solvay & Cie's traditional salt supplier, took over existing Leblanc

---

[8] For parts of this section, I draw on Bertrams, Coupain, and Homburg, *Solvay*, chap. 2, and Philippe Mioche, "Solvay à Dombasle (1870–1914)," in Pierre Lamard and Nicolas Stoskopf (eds.), *L'industrie chimique en question*, Paris, Picard, 2010, 195–225.

manufacturers. Customers became producers, and producers became their own customers – everything was topsy-turvy. Litigations followed suit, and what had hitherto been a (rough) business competition dispute took an unpleasant political turn. In 1884, the general counselor and member of the French trust of salt producers declared in court that "this industry [Solvay & Cie] is only constituted by foreigners, who only hire foreigners and never miss the opportunity to fire the few French workers active at the factory." Solvay & Cie's lawyer vigorously rebutted the argument: out of a thousand workers, 821 were French, 130 came from Alsace-Lorraine, and only 31 were Belgian (mostly employees). Ernest Solvay fulminated against these xenophobic tendencies: "We have always acted [in France] as French industrialists... we will carry on this way."[9] After the First World War, the managing partner and former plant manager of Dombasle, Edouard Hannon, described the real impact of nationalism on the conduct of Solvay & Cie's business. As he argued, "xenophobia was the mere outcome of blinded crowds exerting no kind of significant influence, perhaps with the exception of China" (he referred to the Boxer Rebellion around 1900). Overall, Hannon praised the action of the French "public administrations which have never been systematically antagonistic to our Company; quite the contrary, in fact, for they were more inclined to lean toward the welfare of their country than a fraction of irresponsible and jealous public opinion."[10]

Willing to distinguish between business and politics in France as elsewhere, the Solvays gave priority to industrial agreements over court decisions. Generally, a truce took place only after Solvay & Cie would pursue a policy of price reductions, which would bring the most enduring competitor to its knees. For the French market, when the overarching agreement was eventually concluded in 1896, Solvay & Cie had reached a dominant position, in which it could make proposals that nobody could possibly turn down.

Things were slightly different in Germany. As has already been noted, the political climate was one of euphoria after the Franco-Prussian War, but many things remained to be done to achieve the political integration of the various German "states" into a single

---

[9] ACS, "La politique commerciale des frères Solvay," 28 July 1938, 7.
[10] ACS, Edouard Hannon. "La politique de Solvay & Cie," 31 October 1926, 3.

German Empire. To do so, Bismarck took over the movement toward German unification that had started in the business context with the customs union – that is, the famous Zollverein – back in the early 1830s. Bismarck's general strategy, which was known by the ambiguous term "imperialistic nationalism,"[11] consisted, among other aspects, of rapid industrialization. Despite its appalling social side effects, this gospel directed at heavy industry turned into a stunning triumph. Between 1880 and 1913, the German Empire outpaced its eternal rival, the British Empire, in terms of industrial exports. The output of coal had grown fourfold in thirty years' time, and the production of steel had grown tenfold. Shortly before the Great War, German steel manufacturers produced three times more steel than their British counterparts,[12] but this anticipated version of the German *Wirtschaftswunder* (economic miracle) also affected the more innovative science-based industries, such as chemicals and electrical engineering. Thus, the German Empire had two industrial revolutions for the price of one.

Aware of these impressive social, economic, and political transformations, Ernest Solvay's expectations in Germany were presumably high.[13] Reality, however, was cruelly disappointing. Contrary to the French situation and the British case, the latter of which is discussed subsequently, initial obstacles did not stem from the organization of Leblanc competitors. They came from a chain of bureaucratic and administrative constraints. Solvay & Cie's first intention to erect a factory in Sarralble, located in the German territory of Lorraine, was made official in 1874. The situation dragged on until 1885. An alternative was sought in the meantime and consisted of taking over a soda plant at Wyhlen, near Basel. Here again, however, it took more than six years (1874–80) and a considerable amount of money and effort to have the factory on stream. Solvay & Cie's development in Germany seemed to be

---

[11] Heinrich A. Winkler, *Der lange Weg nach Westen, Band 1. Deutsche Geschichte vom Ende des Alten Reiches bis zum Untergang der Weimarer Republik*, München, C. H. Beck, 2000, 217.
[12] Volker R. Berghahn, *Modern Germany. Society, Economy and Politics in the Twentieth Century*, Cambridge, Cambridge University Press, 1982, 1, 3, 260; Hobsbawm, *Age of Empire*, 47.
[13] Unless otherwise noted, Solvay-related information comes from Bertrams, Coupain, and Homburg, *Solvay*, chap. 2.

Figure 3.3. Workers at the Bernburg plant, 1903. (Solvay Archives)

doomed to follow the quest of the Holy Grail. Just as in Richard Wagner's contemporary operas, however, this did not take into account the sudden rescue of an outside force. Solvay & Cie could count on its own modern knight but, unlike Percival, he really existed – his name was Carl Wessel (1842–1912). As soon as Wessel was hired in 1880, the pace of development accelerated. Shorty after the location for a new soda ash plant had been secured in 1881 at Bernburg, along the River Saale in the Duchy of Anhalt, the building permit was granted. Production started in 1883 and soon reached two-thirds of the output capacity of Dombasle. With Bernburg thriving, Solvay & Cie's "German problem" was solved.

Confident with this achievement in Wilhelmine Germany, Solvay was now staring at the neighboring empire dominated by the Habsburg monarchy. The multicultural empire of Austria-Hungary stretched from Bohemia to Transylvania; it was a cocktail of different nationalities whose aspiration for sovereignty it could precariously maintain by granting a large cultural, religious, and linguistic autonomy.[14] But from a strictly economic standpoint,

---

[14] A. J. P. Taylor, *The Habsburg Monarchy, 1809–1918*, Chicago, The University of Chicago Press, 1948, 156–69; Hobsbawm, *Nations and Nationalism*, 99–100.

the potential was huge. After the ammonia-soda process had been rewarded at the 1873 World Exhibition in Vienna, it was expected that the expansion of Solvay & Cie in the so-called Mitteleuropa was a matter of months. In fact, it took ten years before the Solvay-Werke joint partnership could be established with Austria-Hungary's major chemical company, the Verein Aussig. Although its production capacity was slow to take off, the soda plant built at Ebensee in 1885 paved the way to the large consuming market of southeast Europe.

The impressive industrial breakthrough of Solvay & Cie in the German Empire prompted a more profound change, which touched on the whole structure of the Solvay group. With plants in Bernburg, Wyhlen, and Sarralbe, the current performance and future capacities of Solvay's industrial nexus in Germany and Austria-Hungary were such that they elicited a discussion about organizational matters. Carl Wessel came with the suggestion to set up a limited company that would encapsulate Solvay & Cie's facilities in Germany. It was eventually incorporated under the name Deutsche Solvay Werke (DSW) in 1885, and the capital was largely controlled by Solvay & Cie (some 84 percent altogether). Among the remaining shareholders, an industrialist of German descent based in Britain and an English businessman each held 2 percent of DSW. They were Ludwig Mond and John Brunner. Their capacity as shareholders of DSW was nothing but a remote echo of their influential role in the development of Solvay & Cie in Europe and the world at large.

STRETCHING OVER THE CHANNEL

By the late 1860s, word had spread among chemistry experts that a new process for the manufacturing of soda ash had been designed by a Belgian inventor. Unlike Ernest Solvay at the time of his "discovery," these chemical experts knew about the soda-ammonia reaction and the successive failed attempts for its commercialization. Ludwig Mond (1839–1909) was among these experts. After his time as a chemistry student at several leading Germany universities, he began his professional career in a soda ash plant near Cassel running the Leblanc process. There, he was engaged in various sorts of applied research, which led him to file a patent in 1861 for the invention of a new method to secure sulfur from calcium sulfide waste. This enabled him to travel, notably in Great Britain. In doing so, Mond followed the path of many of his talented compatriots searching for a better position, whether in academia or in industry, than the ones available in Germany at the time. Britain, as a result, was filled with German chemists. August Wilhelm Hofmann had been a pioneer when he was appointed the first-ever director of the Royal College of Chemistry in London in 1845.[15] Likewise, Heinrich Caro made some of its most innovative breakthroughs in the artificial dye industry while working in a company located in Manchester in the early 1860s. But the tide was about to change. Caro returned to Germany in 1866 and became research leader of BASF a few years later.[16] Even Hofmann went back to his home country in 1865, first as a professor at the University of Bonn, then as head of the prestigious Department of Chemistry at the Friedrich-Wilhelms University in Berlin. Both of them were sowing the seeds of the rising science-based economy that would represent the industrial heyday of the German Empire.[17]

Instead of returning to Germany, Ludwig Mond decided to visit Belgium in 1872. There, he insisted to have a guided tour of the Couillet plant facilities. Ernest Solvay reluctantly agreed, and this proved a very good call indeed. Claiming that Mond and Solvay's feelings for each other were that of "friendship at first sight" would seem worthy of an airport romance novel. Nonetheless, it cannot be denied that their personal relations went beyond the traditional borders assigned to business collaboration. They belonged to the same generation and shared many characteristics revolving around innovation and entrepreneurship. Solvay thought highly of Mond's extensive chemical knowledge; the latter was most certainly impressed by the former's vision and determination. However, it took more than mutual admiration to start up their

---

[15] Robert Bud and Gerrylynn K. Roberts, *Science versus Practice. Chemistry in Victorian Britain*, Manchester, Manchester University Press, 1984, 53–75.
[16] Wolfgang von Hippel, "Becoming a Global Corporation – BASF between 1865 and 1900," in Werner Abelshauser, Wolfgang von Hippel, Jeffrey A. Johnson, and Raymond G. Stokes (eds.), *German Industry and Global Enterprise. BASF: The History of a Company*, Cambridge, Cambridge University Press, 2004, 23–6.
[17] Bensaude-Vincent and Stengers, *Histoire de la chimie*, 234–9.

business venture. Underlying Mond and Solvay's partnership was the conviction that the sum of their efforts would lead to mutual benefit. In this sense, Mond's role would not be limited to his activity as promoter of the Solvay technique; he turned out to be instrumental in advising Solvay to carry out a relevant cross-national patenting strategy as well. And he also constantly fostered the technological improvement of the ammonia-soda process. For instance, his obsession with preventing the loss of expensive raw materials was summed up in a phrase he once uttered: "If you can smell ammonia, you are wasting ammonia."[18]

Developing the ammonia-soda process in Britain, the playground of the Leblanc manufacturers, required a well-established strategy. Various joint business options were considered involving also Solvay & Cie's silent partner, Eudore Pirmez. In the end, however, it came down to a rather simple formula: A separate British-based company was to be granted favorable license rights for the production of soda ash using the Solvay process. In essence, for each ton of soda ash produced by Ludwig Mond and his associate, John Brunner, a fixed licensing fee of 8 shillings was paid directly to Solvay & Cie. Brunner, Mond Co. was founded in February 1873 with a first plant built in Northwich the next year. The Leblanc manufacturers might at first have looked at the arrival of this isolated competitor with amused superiority. Less than seven years later, however, overconfidence had given way to anxiety. In 1881, Brunner, Mond Co. was transformed into a limited company, a corporation in which the liability of subscribers is limited to what they have invested. The newly registered company increased its authorized capital and decided to transfer one-fifth of its shares to Solvay & Cie. In exchange, the Belgian company pledged to abolish the licensing fee and other technological and market pretensions on the British territory. Thus, Brunner, Mond Co. was now well equipped to confront the resistance of the Leblanc competitors that, in a desperate and ill-fated assault, formed a coalition to defend its vested interests.[19] This battle lasted until 1926 when, not without irony, Brunner, Mond merged with the leftovers of Leblanc industry and

---

[18] Quoted in W. F. Glasscock, *The Commercial History of the Ammonia-Soda Process*, 1969, 13.
[19] Landes, *Prometheus Unbound*, 247, 273.

Table 3.1. Average annual production of soda ash in Europe (in metric tons)[20]

Years	Leblanc	Ammonia soda (Solvay)
1864–1868	375,000	300
1874–1878	525,000	20,000
1884–1888	515,000	285,000
1894–1898	570,000	680,000
1899–1900	600,000	900,000
1913	125,000	1,860,000

other prominent chemical companies to create the leading conglomerate, Imperial Chemical Industries (ICI). In the meantime, Brunner, Mond had acquired several works that had been started by others: Sandbach (1881), Middlewich (1897), Middlesbrough (1900), Lostock Gralam (1907), and Plumbey (1916). All these sites were located in Cheshire thanks to the concentration of rock salt deposits in this region of northwestern England.

Table 3.1 shows that Ludwig Mond's struggle in the heartlands of the Leblanc industry was pivotal in inverting the balance of power between both processes not only in Britain but in Europe as a whole. This precipitated the extinction of once a prolific industry. As Ernest Solvay put it in 1903, "Born with the [nineteenth] century, the Leblanc process was about to die with that century."[21]

EUROPE IS NOT ENOUGH

Brunner, Mond did not wait to obtain its "imperial" label with ICI in 1926 to start a worldwide expansion, although the arrangements for this expansion had always been negotiated with Solvay & Cie.[22] Among Brunner, Mond's privileged exports destinations was the United States, a country where, contrary to what one might have expected, no soda works of any kind was to be found. From the early 1880s, whether spurred by patriotism or by business interest, an American mining engineer named William B. Cogswell (1834–1921) opposed this British supremacy. After prolonged

---

[20] Glasscock, *Commercial History*, 8.
[21] Ernest Solvay, "Coup d'œil rétrospectif," 600.
[22] This section draws on Bertrams, Coupain, and Homburg, *Solvay*, chap. 2.

talks between Cogswell and Solvay & Cie's representatives, the Solvay Process Company (SPC) was launched in October 1881, with the Belgian company controlling up to half of the shares. The first plant in the United States to manufacture soda ash was located near Syracuse, New York, in a town that would be named after Solvay. Production started in 1884 and would soon thrive. On the other hand, SPC's long-term growth would slowly but surely become eroded by technological mishaps and management mistakes, leaving the door open to unexpected competitors. If, by the late 1880s and early 1890s, European shareholders had nothing to complain about the performance of SPC in a rapidly developing economy, the situation was about to change. The trust between both companies, which rested on strong interpersonal relationships, would time and again be under strain.

What Solvay's American story also revealed was that industrial capitalism "was no longer Eurocentric."[23] Distant parts of the globe were suddenly accessible. The Russian Empire, the largest country in the world, was certainly not the world's most important economy, but it was undoubtedly becoming an emerging market. Of course, political instability undermined the country's development. In 1881, Czar Alexander II did not survive one of the many assassination plots that had been organized against him. His son, Alexander III, succeeded him and was obviously less liberal than his father. Political upheaval, however, was not confined to Russia. It was the rule rather than the exception during these troubled times. One should recall, for instance, that the twentieth president of the United States, James A. Garfield, was shot in March 1881, just four months after his election, and the president of the French Republic, Sadi Carnot, was assassinated by an Italian anarchist in Lyons in June 1894. By and large, however, this lack of security did not stop investors from pursuing their far-reaching activity in the peripheries of the world. Capital investments abroad soared from a nominal value of $9 billion in 1870 to $44 billion in 1913, which stemmed mostly from British and French sources.[24] An increasing part of this transfer of capital concerned not merely loans and equities, but

---

[23] Hobsbawm, *Age of Empire*, 51.
[24] Paul Bairoch, *Commerce extérieur et développement économique de l'Europe au 19ème siècle*, Paris, Mouton-EHESS, 1976, 99–104; Youssef Cassis. *Big Business*.

Figure 3.4. Managers of Solvay & Cie and Brunner, Mond & Co visiting the plant at Syracuse, New York (United States), around 1897. The photograph is taken at the "Castle," headquarters of the Solvay Process Co., located in the middle of the industrial plant. From left to right: Henry R. Cooper, John Wing, J. W. Smith, Frederick Hazard, (below) Nathaniel T. Bacon, (above) Ernest Solvay, (below) Jack Brunner, (above) Edmond Solvay, (below) Fernand van der Straeten, (above) Rowland Hazard (father), (below) Édouard Hannon, (above) Rowland Hazard (son), Edward Trump, William Cogswell. (Solvay Archives)

also represented what has been called "foreign direct investments," which focused on industrial production. Interestingly, while Britain invested massively in its own empire and in the United States, French investment favored emerging markets in Latin America, the Ottoman Empire, and, above all, the Russian Empire, which absorbed a quarter of French foreign direct investment.[25]

Belgium's direct investments abroad were also booming at the time. Belgian-based mining, steel, and railway industries expanded

The European Experience in the Twentieth Century. Oxford: Oxford University Press, 1997, 13–14.
[25] Berger, *Notre première mondialisation*, 27–8.

in Russia, China, and the Independent State of Congo (which was the personal property of King Leopold II until it became a Belgian colony in 1908). Shortly after developing electrical tramways in Russia, Spain, and Egypt, the industrialist Edouard Empain designed and built the underground network in Paris, which was inaugurated for the 1900 World Exhibition (Exposition Universelle).[26] In Russia, observers spoke of a "crusade of Belgian capital" in that country. The circulation of money, engineers, and technology was particularly dense in the regions of Baku and the Donets Basin.[27] Then again, Solvay & Cie's Russian adventure was not due to the initiative of Belgian investors; it sprang from the insistence of a Russian industrialist, Ivan Lubimoff, who sought for, and eventually concluded, a partnership with the Belgian firm. Lubimoff & Cie was founded in 1881, the same year as SPC. A plant was painstakingly erected in Berezniki, near the Ural Mountains, in 1883. The difficult start of the plant and Lubimoff's dubious management prompted the Solvays to engage in the Russian business more actively than they had expected. The Russian company was thus transformed into Lubimoff, Solvay & Cie in 1887, which included outside partners. Another soda works was built in Lissitchansk, in the Donets Basin, in 1891. Contrary to Berezniki, the site had involved the direct participation of Solvay & Cie's higher management (Alfred Solvay and Edouard Hannon), but the executive management was still given to a promising Russian entrepreneur, Vladimir Orloff.[28] Just like Carl Wessel at Deutsche Solvay Werke, Orloff proved an outstanding recruit in conducting Lubimoff, Solvay & Cie on the path toward success.

A MULTILAYERED HEGEMONY

At this stage, we can now reassert that Solvay & Cie's strategy for international expansion was neither uniform nor linear. It owned subsidiaries in Belgium and France, as well as in Spain and Italy,

---

[26] Anne-Myriam Dutrieue, "Edouard Empain," in Ginette Kurgan et al., *Dictionnaire des patrons*, 266–8.
[27] Ginette Kurgan-van Hentenryk, "Banques et entreprises," in Hervé Hasquin (ed.), *La Wallonie, le pays et les hommes* (Tome II). Bruxelles: Renaissance du Livre, 1975, 40–43.
[28] Michel Accarain, *La société Lubimoff, Solvay & Cie, par Wladimir Orlow*, Louvain-la-Neuve, Presses Universitaires de Louvain, 2002, 73–9.

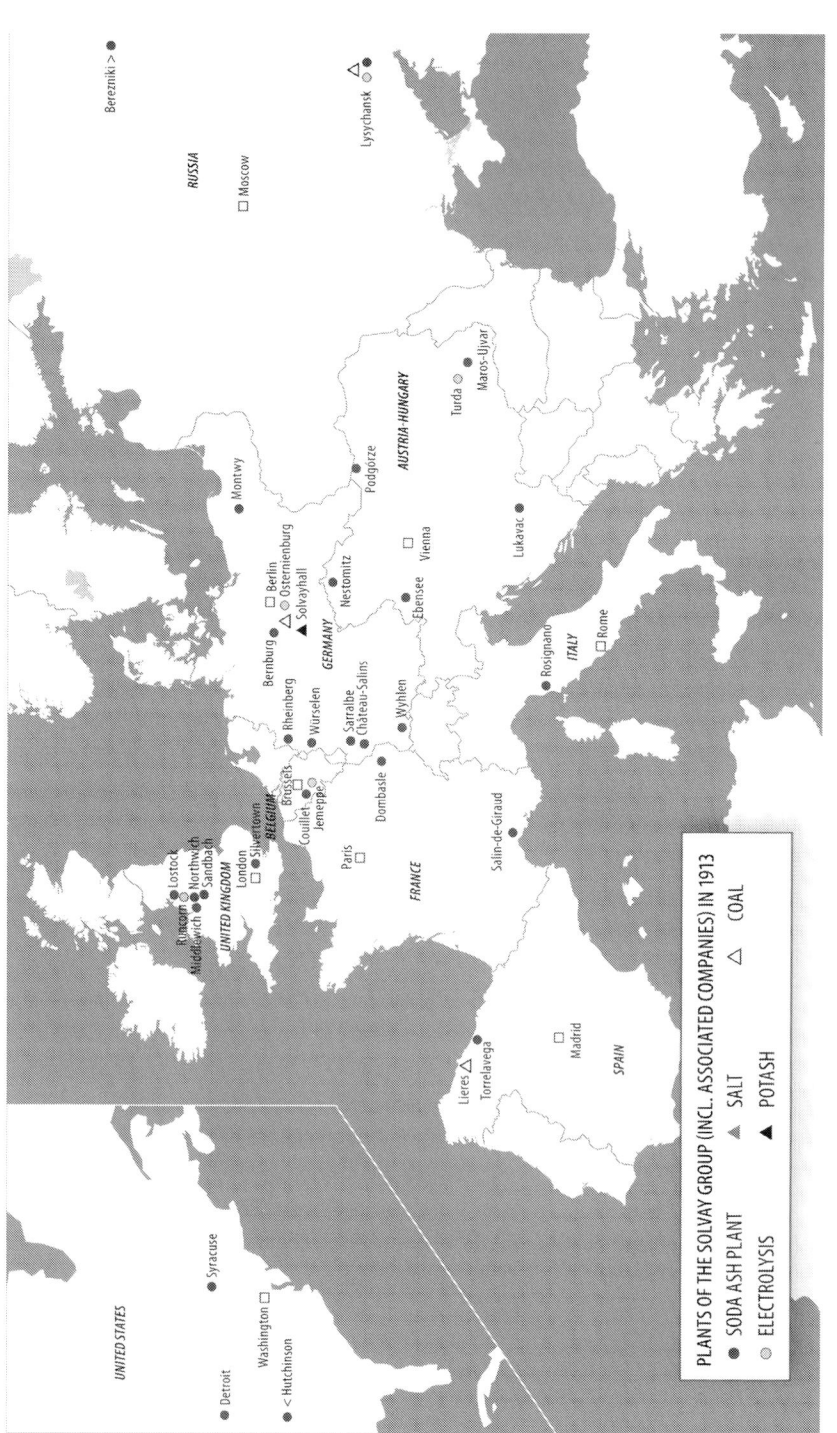

Figure 3.5. Map of the Solvay plants in 1913 (including subsidiaries and associated companies).

after 1900. Elsewhere, it relied on a large network of affiliated companies, which it controlled completely (DSW), as a majority shareholder (SPC), or with more or less equal partners (Lubimoff, Solvay & Cie, Solvay-Verein and, to a lesser extent, Brunner, Mond). This complicated structure was determined by the nature of the foreign market, the amount of investment at stake, and the timing of the negotiations. The Solvay transnational system relied on a mutual exchange of technological know-how, a worldwide division of market shares among partners, and a commitment to suppressing undesirable competition through a wide spectrum of national and cross-national understandings. From an organizational standpoint, Solvay set up a governance model based on strong centralization of strategic decisions mixed with a decentralization of operational management. For political and commercial reasons, the establishment of national works had prevailed over a model based on trade exports alone.[29] Overall, Solvay & Cie had built an industrial empire on the principles of a multilayered hegemony: Its global power rested on the close interconnection of local components and outstanding intermediary actors (Mond, Wessel, Cogswell, Orlow). Therefore, just like any multinational empire, the company endeavored to find the right balance between the loosening grip of federalism and the tighter grip of centralization.

This original blend gave way to a successful achievement. At the eve of the First World War, while celebrating its fiftieth anniversary, the Solvay group had reached its objectives beyond expectations. In 1913, Solvay & Cie and its subsidiaries was the largest chemical business group in the world. Its network of thirty-two plants produced 1.9 million tons of alkali and employed 25,000 people. As Nicolas Coupain put it, "The history of this first half-century of existence is one of immense success of an enterprise that was able to take astute advantage of a technical and commercial opportunity, at the international level."[30] How many, among them, could ever have imagined that the underlying dynamics on which this global expansion had been founded could implode and give way to a major cataclysm?

[29] ACS, "La politique commerciale des frères Solvay," 28 July 1938, 6–7.
[30] Bertrams, Coupain, and Homburg, *Solvay*, 143.

Figure 4.1. British soldiers digging a trench while wearing respirators to guard against fumes from bursting shells. The Germans first used poison gas at Ypres on 22 April 1915. (© Illustrated London News Ltd / Mary Evans Picture Library)

4

World War I and the Collapse of the International Order

> In this autumn of 1919, in which I write, we are at the dead season of our fortunes. The reaction from the exertions, the fears, and the sufferings of the past five years is at its height. Our power of feeling or caring beyond the immediate questions of our own material well-being is temporarily eclipsed. The greatest events outside our own direct experience and the most dreadful anticipations cannot move us.... We have been moved already beyond endurance, and need rest. Never in the lifetime of men now living has the universal element in the soul of man burnt so dimly.
>
> John Maynard Keynes[1]

A DIVE INTO THE DARK

On 12 February 2012, news came that Florence Beatrice Green passed away at the respectable age of 110. More than being one of the oldest persons in the United Kingdom, she was the last known surviving veteran of the First World War. She had briefly joined the Women's Royal Air Force as a stewardess in September 1918.[2] With the death of World War I's last eyewitness, the physical link with this bloodshed will be irremediably lost. What will remain is the traumatic memory left by this unprecedented catastrophe and the legacy of what is still known as the "Great War." To what

---

[1] John Maynard Keynes, *The Economic Consequences of the Peace*, New York, Harcourt, Brace and Howe, 1920, 296–7.
[2] *The Guardian*, 12 February 2012.

extent, however, people today can picture the magnitude of the drama is rather unclear, for the massacre turned Europe into the world's largest graveyard. France lost 20 percent of men of military age, Britain never saw half a million of its soldiers under the age of thirty return home, the Russian Empire suffered almost 2 million casualties, and the German Empire and Austria-Hungary lost some 2 million and 1.1 million combatants, respectively. These morbid figures account for military deaths, although the amount of civilian casualties, mostly due to malnutrition and other diseases, reached unrivaled highs.[3] The conflict's most salient characteristics were that it was a "total war" and the first truly *world* war, stretching within and even beyond Europe's imperial domain.[4]

Although it was often depicted, with some truth, as a European Civil War, the conflict involved all major powers. The prewar system of security-based alliances produced a simplistic world order divided into two opposing military blocks – the Entente Powers and the Central Powers – mitigated by neutral countries. The Entente, or Allied, forces were Great Britain, France, and Russia; they opposed Germany, Austria-Hungary, and Italy, gathered around the Triple Alliance. Italy, however, stalled its participation and ultimately fought for the Allies. All other countries were initially neutral, and only two (Romania in 1916 and the United States in 1917) renounced neutrality and joined the Allies. For the multinational Solvay group, this international clash had clear dramatic consequences. The headquarters of Solvay & Cie were based in Belgium, the neutrality of which was violated as its territory was occupied almost entirely by German troops. Solvay & Cie owned plants in one of the founding Allied members (France), in a former Central Power that became a neo-Allied country (Italy), and in a neutral country (Spain). But there was worse news than this: its foreign subsidiaries were directly confronting each other. Deutsche Solvay Werke and Austria-Hungary's Solvay-Werke were engaged on one side, whereas Brunner, Mond Co., Lubimoff, Solvay & Cie, and Solvay Process Company were on the other. The Solvay group was literally caught up in the war.

[3] Niall Ferguson, *The Pity of War*, London, Penguin, 1998, 293–5.
[4] Eric Hobsbawm, *Age of Extremes. The Short Twentieth Century, 1914–1991*, London: Abacus, 1994, 23–4.

Warfare was a mixture between traditional and new methods. The nature of fighting was dominated by the trench system in the confined areas of the so-called Western Front, by a more mobile form of combat in the larger and more diversified territories involved of the Eastern Front. Relatively new techniques of warfare such as armored vehicles (codenamed "tanks"), machine guns, and, the most controversial of all, poison gas, were brought onto the battlefield. The use of chemical weapons during the war, however ineffective and marginal they eventually proved to be, had stirred huge debates within military, political, and scientific circles.[5] Despite the fact that the First World War has sometimes been perceived as a "chemists' war," traditional weapons largely outweighed new technology. The bulk of combatant casualties died from heavy artillery fire or from small arms used in combat rather than from chemical weapons. Furthermore, the tanks in operation were negligible compared with the millions of horses and mules used (and killed) for the transport of troops and equipment.[6] All of this induced a climate of desolation and instilled a sense of despair among the survivors. As the economist John Maynard Keynes put it with a combination of lyrical style and morbid verve: "A visitor to the salient [of Ypres] early in November 1918, when a few German bodies still added a touch of realism and human horror, and the great struggle was not yet certainly ended, could feel there, as nowhere else, the present outrage of war, and at the same time the tragic and sentimental purification which to the future will in some degree transform its harshness."[7]

"NEVER WAS SO MUCH OWED BY SO MANY TO SO FEW"

The dark side of humanity largely eclipsed its brightness in these harsh times. Nevertheless, a few attempts did somehow restore the balance. Winston Churchill's famous phrase about the Battle of Britain in August 1940 – "Never was so much owed by so many

---

[5] Abelshauser et al., *BASF: The History of a Company*, 165; Roy Mac Leod, "The Chemists Go to War: The Mobilization of Civilian Chemists and the British War Effort, 1914–1918," *Annals of Science*, 50 (5) (1993): 455–81.

[6] David Edgerton, *The Shock of the Old. Technology and Global History since 1900*. Oxford, Oxford University Press, 2007, 34–5, 143.

[7] Keynes, *Economic Consequences*, 95 (n. 80).

to so few" – aptly suits a particular episode of the First World War – namely, an unexpected humanitarian operation in Belgium. Links with Solvay, the family and the company, will become obvious to the reader. It must be recalled that Belgium, whose neutrality had been guaranteed by international treaties in 1831 and 1839, was invaded by German troops on 4 August 1914. After an unexpected resistance around Liege, the German forces were able to sweep through the country provoking a thunder of violence against civilians, which was later exploited by Allied propaganda and German counter-propaganda.[8] As the war turned into a static warfare, Belgium became the laboratory of foreign occupation. For the Belgian citizens who did not flee, the priority was to prevent the country from becoming a laboratory of mass starvation. Belgium only produced one-sixth of the foodstuff it consumed; stocks of grain, sugar, and rice could hold for a maximum of two to three weeks. Many were appalled by the bleak prospect of this situation, but initiatives to counter it came from two industrialists in particular, Ernest Solvay and Dannie Heineman, both of whom acted in close partnership with the mayor of Brussels, Adolphe Max. An American-born engineer with a degree from the Technische Hochschule of Hannover, Heineman (1872–1962) was the head of the multinational electrical holding group, Sofina, and a rising star in business milieus. Through their respective social networks, Solvay and Heineman managed to set up a committee to organize the supply of food and relief goods for the inhabitants of Brussels. In a few weeks' time, the number of requests was so huge that they were easily convinced to extend their organization to the whole territory of Belgium. The National Committee for Relief and Food was created in mid-September 1914 under these circumstances.[9] Its designated honorary chairman was Ernest Solvay, and the head of the executive committee was Emile Francqui (1863–1935), an impressive financier and negotiator then a director of Belgium's most powerful financial holding group, the Société Générale de Belgique.

---

[8] John Horne and Alan Kramer, *German Atrocities, 1914: A History of Denial.* Hew Haven, CT, Yale University Press, 2001.
[9] Liane Ranieri, *Dannie Heineman. Un destin singulier, 1872–1962,* Bruxelles, Racine, 2005, 79–84.

Figure 4.2. General meeting of the Comité National de Secours et d'Alimentation. This private organization chaired by Ernest Solvay acted as a "second Belgian government" during the First World War. It organized the relief of a starving population affected by the conflict and the occupation. (Painting by Henri Lemaire, undated)

Ernest Solvay was not the only member of the family to be involved in the National Committee. An important figure in this respect was his grandson-in-law, Emmanuel Janssen (1879–1954), who had been recruited as General Secretary of Solvay & Cie shortly before the war. Janssen held the key responsibility of chief of the National Committee's *Secours* (benevolence) Department and was as such at the forefront of the negotiations with the German occupying forces. Alfred Solvay's son, Louis, also played a role in the National Committee, as did other employees from Solvay & Cie's administration.

The National Committee took an international turn when, by 19 October 1914, Francqui went to London to convince the British administration to release the food supplies to Belgium. There, he met an old acquaintance, American Herbert Hoover (1874–1964). Less than fifteen years later, on 4 March 1929, the tall and charismatic engineer would be sworn in as the thirty-first president of the United States. In October 1914, however, Hoover, who was a wealthy and successful industrial consultant specialized in mining investment, was busy organizing the repatriation of American citizens stranded on the European continent. It did not take long for Francqui to convince Hoover, who had once been his competitor in China, to take part to a joint relief program. Hoover, after all, had rich experience putting together humanitarian aid. As to the Belgians, they were desperately in search of an American liaison officer whose official neutrality could guarantee the regular shipping of food and goods. It is in that context of wartime emergency that the Committee for Relief in Belgium took shape.[10] Despite occasional rivalries, Francqui's National Committee and Hoover's Committee for Relief, together with the active support of two other neutral powers (Spain and the Netherlands), were able to effectively and successfully set up what is probably one of the first large-scale humanitarian operations in history. Its work as a food supplier, funded by private and public sources, saved the Belgian population, as well as a significant number of people from the northern regions of France, from malnutrition and starvation. The scope of its activity was also political, however, to an extent that goes well beyond the practice of current humanitarian organizations. Fulfilling the mission of public interest, the National Committee progressively grew out as a kind of "second Government" that, no wonder, was not to the liking of German authorities.

The operation's legacy was immense and truly original. As it happened, when the war was over, Francqui and Hoover decided to give the balance of both committees to the building of scientific research facilities in Belgian universities and to endow new institutions dedicated to science and academic exchange.[11] This exceptional bequest paved the way for several other philanthropic initiatives aimed at fostering academic science. Between 1922 and 1924, Ernest Solvay's successors held the reins of the movement by donating the Free University of Brussels some six million Belgian Francs.[12]

---

[10] George Nash, *The Life of Herbert Hoover, Vol. 2: The Humanitarian, 1914–1917*, New York-London, Norton, 1988, 26–9.

[11] Kenneth Bertrams, "De l'action humanitaire à la recherche scientifique: Belgique, 1914–1930," in Ludovic Tournès (ed.), *L'argent de l'influence. Les fondations américaines et leurs réseaux européens*, Paris, Autrement, 2010, 43–61.

[12] Andrée Despy-Meyer and Valérie Montens, "Le mécénat des frères Ernest et Alfred Solvay," in Despy-Meyer and Devriese, eds. *Ernest Solvay*, 243–4.

ORGANIZING THE ECONOMIC MOBILIZATION

Despite some ongoing activity within the neutral countries, wartime antagonism had crushed the prewar values of socialism, liberalism, pacifism, all of which, in their own way, were grounded in a shared commitment toward the making of an international world order. An efficient counterforce to the rising movements of nationalism before 1914, internationalism was one of the war's first casualties. Nationalist and patriotic feelings now ran unleashed without any serious challenge. In terms of political development, military strategy, or economic management, this contributed to legitimate the emergence of a stronger, expanded State. New ministries, bureaucracies, and public agencies, absorbing an ever-growing number of civil servants, suddenly came into the administrative landscape of countries at war. Industrialists were naturally asked to join the war effort led by the State. It prompted them increasingly to opt for original forms of joint State-business associations. This represented the hybrid organizational matrix of the wartime economic drive. Its main initiators were dynamic and surprisingly diverse men such as the socialist politician Albert Thomas and the conservative statesman Georges Clémenceau in France, the liberal leader David Lloyd George in Great Britain, the financier Bernard Baruch in the United States, and the industrialist Walther Rathenau in Germany.[13]

From a business perspective, the war economy yielded mixed results. On one hand, the loss of markets, the induction of significant segments of the workforce, the shortage of feedstock, and various military constraints hindered not only the development but the very practice of industry itself. On the other hand, many companies were more than resilient in these difficult times and did well, especially the largest combines. Enhanced by the State-sponsored environment, industrial mobilization, and far-reaching patriotic aims, the French carmaker Renault became the country's most important company with a workforce soaring from 4,400 in 1914 to 22,000 by 1918.

---

[13] Gerd Hardach, "Industrial Mobilization in 1914–1918: Production, Planning, and Ideology," in Patrick Fridenson (ed.), *The French Home Front, 1914–1918*, Providence-Oxford, Berg Publishers, 1992, p. 65; Gerald Feldman, *Army, Industry, and Labor in Germany, 1914–1918*, Princeton, Princeton University Press, 1966, 46–51.

In the United States, the chemical firm Du Pont grew out of the war as one of the world's most profitable enterprises. Its turnover for the single year 1916 exceeded the combined totals of the years 1902–1914.[14] National legislations favored the shift from soft economic concentration to full-fledged industrial consortia or cartels. In Germany, Bayer Company's leader, Carl Duisberg, inspired the creation of an industrial pool of chemical companies that would serve as the basis for the establishment of the giant I.G. Farben chemical conglomerate in the 1920s. The same profit-pooling design had been implemented in the steel industry in 1915 with the Thyssen concern at its core. It also led ten years later to an enormous amalgamation, the Vereinigte Stahlwerke (United Steel).[15] Altogether, the experience of the First World War proved decisive in framing big business ventures.

INTO THE WAR ECONOMY

How could one describe Solvay & Cie's business activity in this context?[16] First, it must be said that the outbreak of the war and the subsequent occupation of Belgium posed a serious challenge to the otherwise centralized management system of the company. Reactions were immediate: while Ernest Solvay's eldest son, Armand, settled in Biarritz to carry out some of the world organization of the industry, the company's communication division was transferred from occupied Brussels to neutral Rotterdam. In terms of industrial output, then, the situation varied enormously from country to

---

[14] Richard Vinen, *A History in Fragments. Europe in the Twentieth Century*, London, Little Brown, 2000, 52–3; Patrick Fridenson, *Histoire des usines Renault, Vol. 1*, Paris, Seuil, 1972; David A. Hounshell and John K. Smith, *Science and Corporate Strategy: Du Pont R&D, 1902–1980*, Cambridge, Cambridge University Press, 1988, 76.

[15] Robert O. Paxton, "The calcium carbide case and the decriminalization of industrial ententes in France, 1915-1926," in Fridenson, *French Home Front*, 153–80; Jeffrey A. Johnson, "Power of Synthesis," 171–3; Feldman, *Army*, 470; Jeffrey Fear, *Organizing Control. August Thyssen and the Construction of German Corporate Management*, Cambridge, MA, Harvard University Press, 2005, 432–43.

[16] Unless indicated, the next two sections draw on Bertrams, Coupain, and Homburg, *Solvay*, chap. 7.

Figure 4.3. Results of the bombing of the Château-Salins soda ash plant on 24 July 1917. This was one of the few major destructions Solvay experienced during World War I. After the transfer of Alsace-Lorraine from Germany to France, the plant was transferred from DSW to Solvay & Cie. (Solvay Archives)

country, and even on a local scale as far as Belgium was concerned. In contrast to Couillet plant, which resumed its activity after a ten-month interruption following the invasion of German troops, the main production lines of Jemeppe plant remained closed during the entire course of the war. Because of the scarcity of raw materials, the production capacity of Couillet until 1920 was 15 to 20 percent lower than before the war. Then again, the fragile political climate urged Belgian industrialists to maneuver with extreme caution given the pressures exerted by the German occupying authorities, on the one hand, and the legal order that would eventually prevail once the war was over, on the other. More concretely said, there was a thin, red line to draw between the threat of wartime sequestration and the postwar accusation of economic collaboration.

In nonoccupied countries, by contrast, Solvay & Cie's general production followed the hectic impulse determined by the war economy. Although soda ash was not strictly speaking a strategic war-related product, it remained a product of mass consumption during the war, especially with regard to the deteriorating conditions of hygiene. As to Solvay's other flagship product, caustic soda, it was instrumental for the manufacturing of chemical weapons. As a result, the production of soda ash equalled prewar levels at Dombasle and Giraud plants, whereas the output of caustic soda increased twofold. This stemmed largely from the demand created by the State-backed consortium system that was particularly strong in the chemical industry. The French government wished to seize the opportunity of the war to launch a single national industrial "champion" revolving around companies such as Saint-Gobain and Kuhlmann. A National Dyes Company indeed came to fruition. Although the principle of private ownership was maintained, this company represented an important breakthrough from the State in its attempt to reorganize the chemical industry. Nevertheless, it was true that "French chemical companies emerged from the war greatly enriched, with their independence virtually intact."[17] Likewise, this visible upward trend was also observed in Spain, a neutral country, where Solvay & Cie had erected its soda ash plant at Torrelavega in 1908. Alternatively, the production process of Solvay's Italian factory, which was put on stream at Rosignano in 1913, stood mostly still during the time of the war.

On the German side, the creation of new economic agencies, like the Raw Materials Section set up in August 1914 under Walther Rathenau's ingenious supervision, had enduring consequences on the wartime economy up to the interwar. The organization gave way to a complex and centralized system of economic policy but, at the same time, it generated considerable benefit to corporations providing raw materials.[18] As a result, Deutsche Solvay Werke (DSW) walked in the footsteps of the German chemical industry as a whole. With plants located in ten locations across the German Empire, there were evidently some exceptions. For instance, the production of caustic soda at Sarralbe, Würselen, and Osternienburg

---

[17] John F. Godfrey, *Capitalism at War. Industrial Policy and Bureaucracy in France, 1914–1918*, Leamington Spa, Berg Publishers, 1987, 157–80 (at p. 180).
[18] Feldman, *Army*, 45–52.

factories dropped 10 percent each year during the war. However, this was nothing compared with the harsh times to come. Meanwhile, Germany's military partner among the Central Powers, the crumbling Austria-Hungary, showed signs that anticipated the postwar downward trend. There, the plants jointly run by Solvay and Verein Aussig seized the momentum of the wartime economy only to a limited extent. Technological breakdowns, mismanagement, and general disorganization put Solvay-Werke under heavy strain and undermined its future development. With some exaggeration, one could say that it paralleled the Habsburg monarchy's current state of decline, placed in striking contrast with its glorious past. Nobody embodied this combination of decline and disillusionment more than the Habsburgs' last emperor, Charles I, who refused to abdicate and attempted to organize the restoration while in exile in Madeira, carrying with him "the last threads of the Habsburg shroud."[19]

THE GRIM POSTWAR (OR THE PURSUIT OF WAR BY OTHER MEANS)

Handbooks of history remind us that the First World War ended with the capitulation of the Germany army and the subsequent signing of the armistice on 11 November 1918. Looking carefully, however, the period immediately following the cessation of military hostilities hardly resembled what could be called peaceful times. From Helsinki to Budapest, continental Europe was on the brink of a civil war. But this time, the Central Powers were not to blame. The fire was ignited by the Bolshevik's seizure of power in Russia in November 1917, which prompted the fall and annihilation of the Czarist Empire and the conclusion of a separate peace treaty with Germany in March 1918. While promoting the spreading of a world revolution in Central and Western Europe, Lenin's Bolsheviks were confronted by counterrevolutionary uprisings in Russia. This resulted in the outbreak of civil wars in Soviet Russia, the Baltic countries, and Finland, as well as much instability elsewhere in

---

[19] Taylor, *Habsburg*, 251.

Europe between 1918 and 1920.[20] Germany, no wonder, was not spared by the waves of social revolution coming from the east. Yet this could barely distract that country from the peacemaking process prepared in the west. At the Paris Peace Conferences opening in January 1919, the former Allied Powers first intended to develop the global design envisioned by U.S. President Woodrow Wilson enshrined in his famous Fourteen Points. It soon appeared, however, that the Conferences legitimated the return of traditional diplomacy dominated by the Great Powers. Moreover, the alleged thrust for a "just peace" that Wilson and others sought gradually shifted toward what Keynes, as British delegate at Versailles, coined a "Carthaginian Peace." Germany, he lamented, was condemned to pay an excessively heavy tribute in terms of reparations. The French delegation, backed by Belgian representatives, thought differently. Imperial Germany, according to them, was the sole guilty party, and the new German Republic ought accordingly to hold the burden of blame. This belief led to the Franco-Belgian occupation of the Ruhr in 1923. The truth lies perhaps somewhere between these two perspectives. The Versailles Treaty did frame the conditions of a victor's peace, but it left Germany largely intact.[21]

In the meantime, the peacemakers gathered in Paris effectively reshaped the geopolitical contours of Europe. With regard to the political and economic context, adapting the old continent to the challenges of the twentieth century was not an easy task. It gave way to a striking paradox. While President Wilson's principle of "self-determination" was overwhelmingly influential in setting the theoretical foundations of a new, ethnically more homogenous Europe, the United States was conspicuous in its absence in organizing this configuration. After the wartime interlude of 1917–18, political isolationism was once again the U.S. foreign policy toward Europe.[22]

[20] Mark Mazower, *Dark Continent. Europe's Twentieth Century*, London, Penguin, 1998, 8–11; Hobsbawm, *Age of Extremes*, 54–71.
[21] Zara Steiner, *The Lights That Failed. European International History, 1919–1933*, Oxford, Oxford University Press, 2005, 15–70.
[22] An opposite view is provided in Patrick O. Cohrs, *The Unfinished Peace after World War I. America, Britain, and the Stabilisation of Europe, 1919–1932*, Cambridge, Cambridge University Press, 2006.

This provided an interesting analogy with the situation felt by Solvay & Cie with its American partner, Solvay Process Company (SPC). During the war, the upper management of SPC organized a series of capital increases, which, by and large, overlooked their major European shareholders. With pure cynicism, the Hazards brothers, acting as executive head SPC, justified the operation by the fact that the war had diverted SPC from its priorities – namely, its domestic market. Armand Solvay was infuriated by this inelegant procedure, which took place at the worst time during the war. He was able to neutralize the operation but could not wipe out its deteriorating effect when it came to trust and confidence among partners. "There is no doubt," he observed, "that the board of SPC is trying to free itself from all European influence."[23] His contemporaries might have felt that he had exaggerated the situation; in fact, he had anticipated the basic tensions of Solvay's Euro-Atlantic dynamics.

RECASTING NATIONS, RESUMING INDUSTRIAL RELATIONS

As soon as the guns felt silent in Western Europe, the *gérants* realized that the Soviet Revolution, not the war, had deeply affected the Solvay group. At various stages of the Russian civil war, both plants had been seized, occupied, and eventually nationalized by the Bolshevists. Although negotiations to recover the Berezniki and Lissitchansk plants started immediately and dragged on for some fifty years, Solvay & Cie resigned itself at the irremediable loss of its Russian factories. By contrast, the war toll had been relatively minor for Solvay & Cie. Physical destruction and the dismantling of industrial facilities, both of which were pronounced in occupied territories, had been largely averted. Only DSW's small soda plant of Château-Salins suffered significant damages. Elsewhere, it was mostly a question of minor repairs. Other preoccupations soon came to the fore, however. Rumors of huge reorganizations of the European map emerged from the Paris Peace Conferences. Empires went to war in 1914, but nation-states would emerge after

---

[23] Quoted in W. J. Reader, *Imperial Chemical Industries: A History, Vol. II: The First Quarter Century, 1926–1952*, Oxford, Oxford University Press, 1970, 292.

it. Much uncertainty surrounded the fate and political legitimacy of these "successor states."[24] It was obvious that the new international order was to have a profound impact on Solvay & Cie's ramifications in Central Eastern Europe. The company's priority, however, concerned the negotiations pertaining to the so-called German question, in which "question" really meant "problem" and for which no satisfying solution seemed to be found. The defeat had plunged the country into political chaos and economic abyss. If the first postwar years gave alarming signs of economic decline across in Europe, nowhere was this more evident than in the nascent Weimar Republic. For what reasons Germany avoided a civil war between the extreme left and the coalition of counterrevolutionary and anti-Republican forces, historians do not know.[25] Nevertheless, the prospect of a quick recovery was even bleaker there than elsewhere in Europe.

The management of DSW could scarcely feel differently. In fact, the members of the *Vorstand*, Emil Gielen and Ernst Eilsberger, reacted as though they personally experienced the *Diktat* of Versailles. When the Sarralbe and Château-Salins plants followed the fate of Alsace-Lorraine and were transferred to France, there was great bitterness. Although both factories became direct subsidiaries of Solvay & Cie, DSW managers felt they had actually *lost* these plants. Unfortunately for them, the dismantling of the German Empire was not confined to the French borders. An analogous geopolitical rationale applied in the east – among other locations, in the Prussian Province of Posen, where DSW's Hohensalza soda plant was located. According to the provisions of the Versailles Treaty, the area now belonged to the Republic of Poland. The village of Hohensalza was from then on officially called Inowrocław (except during the Nazi occupation of 1939–45).[26] Together with Podgórze plant of Galicia, formerly held by Solvay-Werke in Austria-Hungary, the plant was part of a specific Polish group, the Zakłady Solvay w Polsce, created on 19 May 1921. If the partitioning of the

---

[24] R. J. W. Evans, "The Successor States," in Robert Gerwarth, ed., *Twisted Paths. Europe, 1914–1945*, Cambridge, Cambridge University Press, 2007, 210–35.
[25] Berghahn, *Modern Germany*, 80.
[26] The soda plant was located in the village of Mątwy (Montwy in German), eight kilometers south of Inowrocław.

Habsburg Empire along national lines may have given peacemakers some headache, this was no less true at Solvay & Cie. The six soda plants controlled by the Solvay-Werke venture before 1914 were "redistributed" into no less than five nation-states: Ebensee in Austria; Nestomitz in Czechoslovakia; Podgórze in Poland; Lukavac in the Kingdom of Serbs, Croats, and Slovenes (which formally became the Kingdom of Yugoslavia in 1929); and Torda and Ocna-Muresului in Romania.[27] In these new countries, the *gérants* and their Austrian partners had to negotiate carefully to avoid the nationalization of the works. The objective was eventually reached after intense diplomacy with local industrialists, financiers, and, above all, public authorities. The Solvay-Werke partnership, on the other hand, had run its course and was soon replaced by a complex montage of agreements that formed the basis of a community of interests with the Verein Aussig.[28]

"What the war has shown with ample evidence," wrote Edouard Hannon in 1925, "is the fact that countries should no longer be dependent on plants owned by foreign companies for the supply of war-related products. They instead should facilitate the erection of such plants by nationals."[29] The First World War, in other words, had accelerated the transformation of nationalism as an ideological compound into an elaborate policy. Solvay & Cie, as a multinational company, would have to experience how to cope with the rise of various forms of nationalism and the subsequent consequences. It involved a complete reappraisal of the interplay between global and local spheres, as well as a realignment of the relations between business and politics. Another crucial impact of the war concerned the changing perception of Solvay & Cie's key product: soda ash. As

---

[27] The villages of Torda and Ocna-Muresului were formerly located in Transylvania and were referred to as Turda and Maros-Ujvar in Hungarian. Nevertheless, with the advent of "Greater Romania," thanks to the role played by Romania among the Allies, which was converted into favorable clauses in the Treaty of Trianon, Transylvania was transferred to Romania. Hence, the plants that used to be part of the Ungarische Solvay-Werke network were transformed into a single national Romanian company, Uzinele Solvay din Transylvania.
[28] Verein Aussig was the German denomination of what became the Czech Association Spolek pro chemickou a hutní výrobu, based at Ústí Nad Labem (Aussig in German).
[29] Hannon, "Politique de Solvay & Cie," 3.

Figure 4.4. Solvay executives visiting the Wieliczka salt mine, near Krakow, Poland (1921). In the front row (from left), the *gérants* Émile Tournay and Louis Solvay. Next to them is Sigismond Toeplitz, head of the Polish subsidiary Zakłady Solvay. (Solvay Archives)

the *gérant* Emmanuel Janssen put it, the war provoked "the vanishing of the mystery which surrounded the soda industry."[30] That "mystery" had been kept alive for some forty years largely because of the commercial policy the company pursued. The war economy cast alkali products a sharp light. They were no longer invisible, so to speak. As a result, soda ash and caustic soda were now also perceived as something more than products of mass consumption; they became products of general or public interest. The high levels of state intervention after the war had to be seen in this respect. In the end, however, these expressions of the legacy of the First World War represented but two among the many challenges the company was about to face during the interwar.

[30] ACS, Emmanuel Janssen, "Voyage aux Etats-Unis (septembre-octobre 1926)," 28 October 1926, 24.

THE TWILIGHT OF AN ERA

Between 21 and 27 April 1922, chemists gathered at the first Solvay Conference in Chemistry held in Brussels. There had already been three Solvay conferences in the field of physics, but this was the first time that the Solvay experience was dedicated to the discipline that had been so vital to the company's success.[31] The official photograph of the meeting depicts an elderly Ernest Solvay sitting in the front row and holding firm to his cane. He died one month after the picture was taken, on 26 May 1922.

Ernest Solvay's legacy is immense, to be sure. Focusing on his philanthropic action alone, one can only wonder at his achievements when he came to fund scientific research and institutes. The underlying idea was to promote science-in-the-making rather than its most outstanding findings, as the Nobel Prizes would do. In this sense, Solvay's philanthropic action was truly modern. But he was no less active in other fields. His physical and sociological theories sprang from an overarching philosophy that derived from the positive laws of natural science, but he constantly sought to demonstrate their scientific truth through proper experimental methodology.[32] Elected twice as Liberal representative in the Belgian Senate (1892–94 and 1897–1900), he was neither a party affiliate nor a political activist. Yet if his political career was short-lived, the same could not be said of his political influence, commitment, or, better put, his political *vision*. A true supporter of the borderless circulation and promotion of intelligence, Ernest Solvay was convinced by the virtues of internationalism. Hence, he was profoundly wounded when the war broke out. The violation of Belgium's neutrality by the Kaiser's troops affected him on a more personal standpoint. His patriotic feelings grew stronger during the war and, as we have seen, found an appropriate conversion in the organization of the National

---

[31] Ernest Solvay funded the third congress of the International Association of Chemical Societies (IACS) held in Brussels between 19 and 23 September 1913, which coincided with the celebration of the company's fiftieth anniversary. The IACS setting was intended to form the basis of Solvay's International Institute for Chemistry, but the project was preempted by the war. See Brigitte Van Tiggelen and Danielle Fauque, "The Formation of the International Association of Chemical Societies," *Chemistry International*, 34 (1), 2012, 8–11.

[32] For an overview, see Bertrams, Coupain, and Homburg, *Solvay*, chap. 5.

Figure 4.5. The first Solvay Conference of Physics of 1911 gathered the most brilliant minds of the time (e.g., Nernst, Planck, Brillouin, Rubens, Sommerfeld, Lindemann, Lorentz, De Broglie, Knudsen, Warburg, Perrin, Hasenohrl, Wien, Curie, Jeans, Rutherford, Poincaré, Kamerlingh Onnes, Einstein, Langevin, together with Belgian organizers Solvay, Goldschmidt, Herzen, and Hostelet). This picture remains an icon in the history of science. (Picture by Benjamin Couprié)

Committee for Relief and Food. After the war, when Solvay refused to be given a peerage, King Albert I of Belgium appointed him to the honorary political function of ministre d'État in recognition of his humanitarian accomplishment.

It is not a coincidence that Ernest Solvay's ultimate political proposals were sent to the delegates at the Paris Peace Conferences in 1919. One of them concerned the "general solution for the universal problem of social organization," which, in the author's personal style, could be understood in such a way that he endorsed the mechanisms of social justice and peace settlements inspired by Wilsonian idealism.[33] At the same time, however, he was aware that the patterns of internationalism were out of joint. They had collapsed along with the prewar world order. When Belgian and French scientists insisted that their German colleagues should not be invited to participate in the conferences organized by Solvay's International Institutes of Physics and Chemistry, he did not oppose them.[34] The boycott came to an end after the Locarno Treaties of 1925, a period that brought a glimpse of hope as far as Europe's political climate was concerned. Both diplomatic and scientific relations between former belligerents could resume. It was a form of appeasement without reconciliation, but even this was provisional.

---

[33] Jean-François Crombois, "Energétisme et productivisme: la pensée morale, sociale et politique d'Ernest Solvay," in Despy-Meyer and Devriese, eds., *Ernest Solvay*, 220.

[34] Kenneth Bertrams, "Caught-up by Politics? The Solvay Councils on Physics and the Trials of Neutrality," in Rebecka Lettevall, Geert Somsen, and Sven Widmalm (eds.), *The Science, Culture and Politics of Neutrality in Twentieth-Century Europe*, London, Routledge, 2012, 140–58.

Figure 5.1. Solvay Mercury cells at the Jemeppe electrolytic plant, 1910. (Solvay Archives)

5

The Rationalization of the World Chemical Industry

> No Congress of the United States ever assembled, on surveying the state of the Union, has met with a more pleasing prospect than that which appears at the present time.... The great wealth created by our enterprise and industry, and saved by our economy, has had the widest distribution among our own people, and has gone out in a steady stream to serve the charity and the business of the world. The requirements of existence have passed beyond the standard of necessity into the region of luxury. Enlarging production is consumed by an increasing demand at home and an expanding commerce abroad. The country can regard the present with satisfaction and anticipate the future with optimism.
>
> Calvin Coolidge, President of the United States
> State of the Union address, 4 December 1928

BIGGER BUSINESSES, MORE PRODUCTS: A VIEW FROM THE UNITED STATES

In addition to political factors, major economic changes also had a deep impact on postwar societies. The war had caused a huge increase in the production of war-related products, such as propellants and other special chemicals, for which the market prospects were minimal after 1918. Iron, steel, and chemical companies that had functioned on the mass production patterns of wartime economy were suddenly facing the threat of underutilized products and personnel. New market conditions posed a series of challenges to existing businesses. These conditions varied by local circumstances, but they all shared a basic concern: the need to adapt to the economic world of the 1920s. A special mechanism was regularly conveyed in this respect: rationalization. During the postwar years, industrial milieus seemed to have expressed an all-around obsession with various forms of organizational techniques that stemmed from engineering-based rationalization, also called *Americanism*.[1] Of course, Frederick Taylor's so-called scientific management and, later, Henry Ford's mass production techniques were the most widespread and common manifestations of rationalization implemented on the shop floor. But there was also a large-scale version of rationalization, which concerned wider industrial environments rather than factories. It was carried out through diversification and concentration.

While the latter had already appeared at the turn of the twentieth century, the former was a rather new full-fledged industrial strategy. Pioneering in diversification was the U.S. chemical company Du Pont. In a short time span, the family business was transformed from predominantly an explosives manufacturer into a company producing several derivatives from explosives chemistry (paints, dyestuffs, celluloid plastics, films, manmade fibers). But there was more to this. The expansion of its product lines was made possible through a profound reorganization eventually adopted in September 1921, after which Du Pont's management structure became decentralized and "multidivisional." Different divisions were created that enjoyed more autonomy and flexibility with respect to the company's administrative chart. A special emphasis was put on product development through the investment in research facilities.[2] According to the historian Alfred Chandler, big industrial companies that followed a multidivisional policy in the 1920s did so as a means to restore growth and exploit economies of scale and scope.[3] The

---

[1] Mary Nolan, *Visions of Modernity: American Business and the Modernization of Germany*, New York, Oxford University Press, 1994.
[2] Pap Ndiaye, *Nylon and Bombs. Du Pont and the March of Modern America*, Baltimore, Johns Hopkins University Press, 2007, 62–7; Hounshell and Smith, *Science and Corporate Strategy*, 98–110.
[3] Alfred D. Chandler, *Scale and Scope. The Dynamics of Industrial Capitalism*, Cambridge, Harvard University Press, 1990, 40–5.

strategy proved largely successful for Du Pont because it "permitted the company to turn in an impressive profit record even during the years of the Great Depression."[4] Diversification was thus a strategy intending to reconquer the market through the supply of multiple products on the one hand, and to generate cost savings on the other.

Another example, although more modest in scope, was Semet-Solvay Company. When the manager of Solvay Process Company (SPC), Rowland Hazard, acquired the patents of the Semet-Solvay process for the American market in 1895, the new company, which was largely controlled by SPC and American capital, concentrated on the construction of by-product coke ovens. This was the norm: The recovery technology designed by Ernest Solvay's son-in-law, Louis Semet, and commercialized by the *gérant*, Edgard Hulin, proved well adapted for the manufacturing of soda ash using the Solvay process. Yet by the late 1900s, Semet-Solvay Company soon branched out into a large spectrum of coke-oven gas products (benzene, toluene, naphta, etc.). Shortly before the war, Semet-Solvay Company's investments tactics had reached such levels that they almost outweighed those of its holding company, SPC.[5] Interestingly, Semet-Solvay Company's Belgian counterpart, the Société Anonyme des Fours à Coke Semet-Solvay, created as joint-stock company in 1912, would later on follow a different path for diversification. Its strategy in the late 1920s was to foster expansion through mergers and interlocking participations. This was closer to the mechanisms of business concentration.

Just like diversification, the ultimate goals pursued by an industrial concentration were cost reduction and the reordering of competitiveness. But it also permitted an increase of financial capacity. Therefore, concentration frequently appeared as a prelude to diversification. As already noted, regrouping companies into one larger unit was nothing new, especially in the United States. A first wave of mergers had already taken place there in the 1890s. The world pioneer in this respect was John D. Rockefeller's famous Standard

[4] Ibid., 177.
[5] Bertrams, Coupain, and Homburg, *Solvay*, chap. 5.

Oil Trust created in 1882. In the American chemical industry, the most important turn-of-century consolidations led to the setting up of the Barrett Company in 1896, General Chemical in 1899, and Du Pont in 1902. While General Chemical was a merger of eleven producers concentrating mostly on the production of sulfuric acid, the constituents of what became Barrett Company all focused on coal-tar chemicals and their derivatives.[6] Their respective story is important for Solvay & Cie for at least two reasons. The first reason brings us before the war. Back in 1910, Barrett Company, General Chemical, and Semet-Solvay Company joined forces and created Benzol Products, a subsidiary that developed specialized coal-tar intermediates. Then, in 1917, Benzol merged with dye producers to form National Aniline & Chemical. As a result of these amalgamations, the postwar American chemical industry was less fragmented, and so was the market. But this series of mergers only constituted an appetizer compared with what came next – namely, the creation of Allied Chemical & Dye in December 1920. This is the second reason of this Solvay-related evocation.

ALLIED CHEMICAL OR THE MYSTERIOUS MISTER WEBER

If concentrations and mergers were most common in the United States, they became rather en vogue in Europe as well throughout the 1920s. Some were spectacular, but many were not that successful.[7] Part of these relative failures was the difficult integration due to the lack of previous connections. From what was written above, the five companies which formed Allied Chemical & Dye were not unknown to each other when the merger became official. They were General Chemical, the Barrett Company, National Aniline & Chemical, Semet-Solvay Company, and SPC. Some might even claim that Allied was the natural consequence of the business relationships that had taken place previously. The arguments for the 1920 merger were explicitly set forth by the committee of

---

[6] William Haynes, *American Chemical Industry, Vol. VI: The Chemical Companies*, New York, Van Nostrand, 1949, 45–7, 179–81.
[7] Cassis, *Big Business*, 31.

organization: "greater diversification of output and correspondingly greater stability of business; closer adjustment of the production of basic and intermediate materials to the requirements for manufacture and their derivatives; and greater financial strength."[8] The formation of Allied, thus, resulted from an elaborate strategy that aimed at making optimal use of the five companies' resources and personnel. One could guess such strategy was in line with Solvay & Cie's own intentions. This was not the case.

Brussels was largely taken by surprise by the haste with which Allied was formed. When the *gérants* first heard about the project back in 1918, they had favored alternative combinations, the first of which was to arrange an agreement with the rising giant Du Pont, also a family business. Although they mostly disagreed about the formation of Allied, they did not oppose it. Why? A plausible explanation was that the *gérants* saw the merger as an opportunity to improve performance at SPC, for the American subsidiary had been losing market shares for years. The other reason was that they were largely *unable* to oppose it. As it happened, SPC's attitude toward its major shareholder had not been cooperative since the war, to say the least. Besides, SPC certainly did not have the upper hand during negotiations that led to the creation of Allied. Experienced businessmen like the head of General Chemical, William H. Nichols, and, above all, the financier behind National Aniline, Eugen Meyer, were pulling strings. A clear signal came when Meyer managed to impose his successor at the head of the new combine, a professional manager and strong character named Orlando Franklin Weber (1878–1945). Weber's autocratic management style and secretive manner proved instrumental in leading Allied on the way toward business success – and bringing its staff to the brink of a nervous breakdown. His relations with Solvay & Cie were scarcely as stable as the index of the New York Stock Exchange; they fluctuated between polite skepticism and overt confrontation.

Weber's biography is an epic story in itself. The son of a socialist union leader based in Milwaukee, he was a cycling champion

---

[8] Quoted in Chandler, *Scale and Scope*, 179.

Figure 5.2. Solvay owned a 16 percent participation in the powerful American trust Allied Chemical & Dye Corp. at its founding in 1920. Despite the slackening industrial relationship, Solvay remained a foremost shareholder of this combine, which remained in the first ranks of the American chemical industry for some time. In 1963, the central offices of Allied Chemical were housed at 1 Times Square in New York, demonstrating the power of the firm. (Allied Chemical Archives, courtesy of Honeywell)

who became a salesman in the nascent automobile industry. Moving around the United States, he eventually drew the attention of Eugen Meyer, who admired his thirst for action and organizational capabilities. These two were expected to form a successful team at National Aniline in the midst of the war. "Meyer was conceiving and Weber was executing," Emmanuel Janssen later portrayed them. "As soon as Meyer took over some business, he would give Weber full power to reorganize it. That is how Weber progressively took his dictatorial habits from. These were troubled times. One had to react quickly. And all that was needed to earn money in abundance was to have good organizational skills."[9] There was more than just a cultural gap between Weber and the second-generation leaders of the Solvay family business; an ocean separated them. While Emmanuel Janssen showed understanding, his colleagues at the *gérance* almost choked when they heard about Weber's expected salary as CEO. Yet the mysterious manager seemed to be a unanimous choice for the job, and his performance soon supported his unabashed self-confidence. Production costs were dropping, sales and turnover were increasing, and profits were soaring. The impression was characteristic of the booming and speculative times of the so-called roaring twenties. What could shareholders complain about? Holding a bit more than 16 percent of the capital of Allied through its 50 percent holding in SPC, Solvay & Cie appeared relieved while collecting the generous dividends. However, the honeymoon would soon draw to a close.

The first signs of trouble came in 1922.[10] First, Weber was reproached the recklessness of his managerial method. He had fired several members of the middle management as well as skilled technicians. They had opposed his despotic ways of wielding his orders over them. European shareholders were aware of this: They criticized less Weber's sacking technique and motivations than their outcomes. For this, expert personnel had immediately been recruited by direct competitors, all of which were curtailing Allied's market

---

[9] ACS, Emmanuel Janssen, «Voyage aux Etats-Unis (septembre-octobre 1926)», 28 Oct. 1926, 10.
[10] See Bertrams, Coupain, and Homburg, *Solvay*, chap. 9.

Figure 5.3. A rare picture of Orlando Weber (in profile on the right), taken in 1935 after his retirement as chairman of Allied Chemical & Dye Corp. His suspicion of journalists and photographers led *Life* magazine to call him the "mystery man of Wall Street" in a 1938 article. (Photograph by Robert Boyd, *Milwaukee Journal*)

shares. Solvay & Cie and Brunner, Mond feared that the know-how of the soda-ammonia process, which they had managed to keep secret for more than fifty years, was fully disclosed. While Weber downplayed the danger, it was obvious that his management practices sharply contrasted with that of his European counterparts. This was the cornerstone of a wider misconception: Weber expected that shareholders remained silent shareholders, as long as business was thriving, of course. Solvay & Cie, on the contrary, argued that their relations ought to be based on the principles of an industrial partnership. Hence, the *gérants* set the impetus to draft what was called a "tripartite agreement." Signed in July 1922, the document intended to establish a closer exchange of technical information and a reordering of the worldwide commercial positions between Solvay & Cie, Brunner, Mond and SPC. Due to rising mutual suspicion, however, the agreement never came close to application. The issue went well beyond the scope of bilateral Allied-Solvay relationships.

As the chemical industry became increasingly globalized and organized into bigger units, the causes of the tensions lay elsewhere, that is, in the shaky Weimar Republic.

THE SYNTHETIC AMMONIA YEARS (1919–1924)

In the early 1920s, the German economy was still paying the price of war, literally speaking. Beside the colossal amount of reparations, to which the Allied Powers' war debts were added, Germany was forced to submit to technology transfer of important information. The French government used the provisions of the Versailles Treaty to require the German chemical industry to license one of its key wartime technologies: the Haber-Bosch process for the production of synthetic ammonia by nitrogen fixation. This was not surprising: The process was a revolutionary breakthrough for the production of explosives and fertilizers.[11] For the French State, obtaining the patent rights virtually free of charge was a bargain. Of course, the BASF Company, which owned that technology, was not eager to give in so easily to the French *Diktat*. The head of BASF and cofounder of the process, Carl Bosch, wanted to exploit the general interest in this technology and was open to licensing negotiations.[12] For reasons linked mostly to securing the advantages of its own technology, Solvay & Cie was also interested by the Haber-Bosch process. At a time, the Solvay group perceived it as a potential commercial threat. Together with delegates from Brunner, Mond, the *gérants* Edouard Hannon and Emmanuel Janssen had made inspection visits at BASF works at Oppau near Mannheim in August 1919. The production capacity of pure nitrogen was already impressive there but still below the levels planned for BASF's factory in Leuna, south of Bernburg, which was still under construction.

---

[11] Anthony S. Travis, "High Pressure Industrial Chemistry: The First Steps, 1909–1913, and the Impact," in Anthony S. Travis, Harm G. Schröter, Ernst Homburg, and Peter J. T. Morris, eds., *Determinants in the Evolution of the European Chemical Industry, 1900–1939*, Dordrecht, Kluwer, 1998, 8–13.

[12] L. F. Haber, *The Chemical Industry, 1900–1930: International Growth and Technological Change*, Oxford, Clarendon Press, 1971, 194–195; Abelshauser et al., *BASF: The History of a Company*, 183–5.

Solvay & Cie was willing to start negotiations with BASF and, through it, with the eight other chemical companies gathered in the "community of interests" that had been combined in 1916, known as the "little I.G. Farben." There were constraints, however. The ashes of war were still smoking, and the international context was far from stabilized. Collective Inter-Allied initiatives were in the air. On the French side, the postwar patterns of state interventionism prompted the creation of a company intending to exploit the Haber-Bosch process. This had given way to nationalist controversy because an alternative technology was made available at that time by the Frenchman Georges Claude. The French Ministry of Munitions could not care less and moved on with a technique "made in Germany." Solvay & Cie had been invited to join in but had immediately declined the offer; it was reluctant to take part to a State-owned or State-influenced venture. The hunch proved correct: The French State's painstaking efforts led to the erection of a disastrous plant in Toulouse in 1929, which was working at a loss.[13] The British government was also exerting pressure on Brunner, Mond to enter into high-pressure synthesis technology. They finally announced they were building a plant in Billingham for the production of synthetic ammonia running the Haber-Bosch process. BASF was furious: The technology transfer had not been negotiated between companies. Brunner, Mond had obtained information on the know-how through a secret purchase from two Alsatian engineers employed at Oppau during the war.[14] With its British partner opting to go for it alone, Solvay & Cie was given leeway to open discussion with BASF. An initial meeting took place in Aachen in September 1919. Everything was happening fast. The Versailles Treaty had just been signed, and Inter-Allied troops, including Belgian divisions, still occupied the West bank of the Rhine.[15]

By negotiating with Carl Bosch and his counterpart at Bayer, Carl Duisberg, the *gérants*' intentions were twofold. First, they wished

---

[13] Fred Aftalion, *A History of the International Chemical Industry*, Philadelphia, Chemical Heritage Foundation, 2001, 144.
[14] Reader, *Imperial Chemical Industries, Vol. II*, 355–8; Abelshauser et al., *BASF: The History of a Company*, 186.
[15] Margaret Pawley, *The Watch on the Rhine. The Military Occupation of the Rhineland*, London, MacMillan, 2007, 39–41.

to neutralize the potential risk of increased competition in the alkali business. They were quickly reassured on this point: Neither BASF nor Bayer planned to concentrate further on the production of soda-related products. The second objective was more ambitious, albeit less clear. Solvay & Cie sought to engage both companies in an international coalition with Brunner, Mond and Allied Chemical. Synthetic ammonia, in this regard, was a first step toward the establishment of a wider commercial project, but it was not really defined. Bosch and Duisberg nodded respectfully. They neither accepted nor declined the proposal; they merely asked for time to reflect on the matter. After all, no one thought there was any rush. This was not true, however: Solvay & Cie's international partners were expanding as fast as lightning. Relying on previous discussions between General Chemical and SPC, the first American company to start processing the ammonia synthesis, Atmospheric Nitrogen Corporation, was created in 1920 and became part of Allied Chemical.[16] However, it took Orlando Weber until 1927 to erect a nitrogen plant in Hopewell, Virginia, which could scale up the results that had been reached on a small pilot plant. As for Brunner, Mond, its leadership in the British chemical industry was, after several acquisitions, now undisputed. Solvay & Cie feared being left out while giants were being formed. As a result, the *gérants* pushed for a direct understanding between Solvay & Cie on the one hand, and BASF, Bayer, and the remaining members of the "little I.G. Farben" on the other.

Although another wave of meetings took place in 1922, at the time of the "tripartite agreement" between Solvay & Cie, Brunner, Mond, and SPC, the German agreement dragged on for months but was ultimately signed in the summer of 1924. Why did it take so long? Part of the answer points to Orlando Weber. The increasing power gained by Allied Chemical progressively persuaded Bosch and Duisberg to resume conversations with Solvay & Cie. Without a doubt, they were longing for a piece of the American pie and felt Solvay & Cie was an opportune intermediary in this respect. The other part of the answer probably lay in the political and economic context. Germany had stalled its payments of gold and deliveries

---

[16] Travis, "High Pressure," 16.

of coal and timber by late 1922, in breach of war reparations. This prompted French and Belgian troops to dash toward the Rhine and occupy the Ruhr Region's rich mining deposits starting in January 1923. Of course, it did not contribute to restore an environment of international normalcy. Economic chaos in the Weimar Republic followed, spurred by unprecedented levels of hyperinflation (a loaf of bread cost millions of paper marks) and series of uncontrolled strikes. The last French and Belgian soldiers left the Ruhr region by August 1925, in the wake of the Locarno Treaties.[17] The agreement between Solvay & Cie and the "little I.G. Farben" was thus concluded in an oppressive atmosphere. Its content was of utmost importance, however. It consisted of two separate contracts: an exchange of shares between DSW and the German group on the one hand, and a commercial contract based on the respect of previous product lines on the other. It was a significant and successful step for Solvay & Cie in its aim to build a strong alliance in line with international standards. Nonetheless, obstacles would soon interfere.

THE "MAGIC SQUARE" VENTURE THAT NEVER CAME THROUGH (1925–1926)

The deal Solvay & Cie had just reached with BASF and the others was strengthened by a further decision taken in the following months on the German side: the consolidation of the "community of interests" into a single unit. The "little I.G." was transformed into I.G. Farben AG. Bosch had finally won over Duisberg in operating this spectacular merger of eight constituents, which he had envisioned for a long time. Officially incorporated in December 1925, I.G. Farben soon became the largest and most powerful chemical company ever created in Germany.[18] With the partnership between Solvay & Cie and I.G. Farben, all eyes stared across the Atlantic.

[17] Charles H. Feinstein, Peter Temin, and Gianni Toniolo, *The European Economy between the Wars*, Oxford, Oxford University Press, 1997, 36–41; Mazower, *Dark Continent*, 66.
[18] Chandler, *Scale and Scope*, 564–9; Cassis, *Big Business*, 50–1; Abelshauser et al., *BASF: The History of a Company*, 201–5.

What did Orlando Weber have to say? Was he interested in forming a new version of the Triple Entente – quite different, one must admit, from the wartime military alliance? "For the time being," noted the *gérants* in November 1923, "Americans seem to have the vertigo of their power. They have an exaggerated [tendency] to consider themselves as the world's arbitrators."[19] What was obvious, on the other hand, was the growing animosity that emerged between Weber and Brunner, Mond's chairman, Roscoe Brunner, the son of cofounder John Brunner. The stumbling block of the disagreement was difficult to confront because it concerned the opposition of personal styles and cultures, which was related to the so-called special relationship between Britain and the United States. Weber and Bruner's mutual hostility was sharp enough that they kept each other at bay. But the story turned tragic when, in 1923, Roscoe Brunner committed suicide after losing an important lawsuit to Lever Brothers, filed after a breach of contract. He was succeeded as chairman by another family scion, the son of Ludwig Mond, Alfred Mond, who had spent seventeen years in politics with a career climax in 1921–2 when he was appointed minister of health.

Alfred Mond was more favorably disposed to the international alliance that Solvay & Cie pursued than Roscoe Brunner had been. At least he wanted to give it a try. At Solvay & Cie, Emmanuel Janssen was the enthusiastic architect of the alliance. He had relentlessly arranged negotiations with Allied, I.G. Farben, and Brunner, Mond separately. Now he believed the time was ripe for the formation of a "magic square," a fourfold international Allied or I.G. Therefore, he added a bit of drama in his last annual report regarding his mission to the United States: "The situation is serious," he warned his colleagues as he returned in November 1925 from a long stay overseas. He went on by advocating the creation of a four-branch holding venture revolving around Solvay & Cie's foreign partners.[20] Unknown in the equation, however, was Weber's

---

[19] ACS, Minutes of the gérance, 12 November 1923.
[20] ACS, Emmanuel Janssen, "Voyage aux Etats-Unis (octobre–novembre 1925)," November 1925, 17.

attitude with respect to the previous arrangements. While in Europe in January 1926, he claimed that the "tripartite agreement" signed in 1922 was in sharp contradiction to the provisions of the 1890 Sherman Act, a piece of antitrust legislation that limited the formation of cartels and monopolies. This was both unconvincing and hypocritical. It was unconvincing because, paradoxically, the Sherman Act had encouraged, rather than prevented, the merger movement and the harnessing of free-market–based competition. In the steel industry, for instance, the concentration levels in Germany reached the American-style standards only in 1926 with the creation of Vereinigte Stahlwerke.[21] It was hypocritical because, all of a sudden and especially since the end of the war, the Sherman Act was deployed as a legal shield against foreign-based companies organizing in the United States. Such rhetoric legitimated economic nationalism in the name of liberalism.

In view of this, 1926 was undoubtedly one of the most exciting years in the history of the multinational firm. Consider the correspondence exchanged, the meetings held, the discussions organized. When something stood firm one day, such a stance could be shredded to pieces the next. Nobody knew what to expect as far as international agreements were concerned. Eventually, a last-ditch meeting took place at Allied's headquarters in New York City in September 1926. Quite exceptionally, two other *gérants* accompanied Emmanuel Janssen on his trip overseas: Ernest Solvay's grandson, Ernest-John Solvay (1895–1972), and one of Alfred Solvay's son-in-laws, Emile Tournay (1878–1958). They listened carefully to what Weber told them. Invoking the Sherman Act, the general manager wanted "complete freedom of action" for *his* company. This, as they later understood but still could not believe, meant not only the suppression of the "tripartite agreement," but also the severing of all previous contracts determining the relationships between Solvay & Cie and Allied's constituents, SPC and Semet-Solvay Company. This was more than Ernest-John Solvay and Emile Tournay could bear, and they left on the next steamship to Europe. Emmanuel Janssen stayed a while in New York. He did not want to reason with Weber; he simply wished that he would pull

---

[21] Fear, *Organizing Control*, 239–41; Cassis, *Big Business*, 49.

himself together. Allied's CEO, Janssen believed, was both a business genius and an unpredictable actor.[22] In both the theatrical and business realms, however, Orlando Weber had overreacted. It was too late now; the project was dead. Alternate scenarios unfolded, first at Brunner, Mond, then at Solvay & Cie, each with very different outcomes.

ICI, THE BRITISH EMPIRE OF CHEMISTRY

Defining Imperial Chemical Industries (ICI) as a backup plan might sound irreverent toward one of Britain's most ambitious ventures of the twentieth century. As the historian Alfred Chandler noted, ICI "was the first merger in Britain to consolidate major sectors of a basic industry in the American manner."[23] It was nonetheless Alfred Mond's second-best choice and an all-British option, which he had long turned down. As a matter of fact, Weber's hostile attitude proved critical in discarding Mond's ultimate hesitation. ICI thus consisted of an amalgamation of Brunner, Mond with Nobel Industries, itself the result of a series of consolidations that took place shortly after the war, the formerly State-owned British Dyestuffs, and United Alkali, the stronghold of the Leblanc producers. For the latter, the merger was a nice way to draw a veil over an important, yet obsolete process that had enabled the company to enjoy a short-lived renaissance during the war.[24] Alfred Mond became chairman of the new "imperial" combine, and Nobel Industries' general manager, Harry McGowan, was appointed president. Much like Weber, with whom he shared a working-class background, McGowan was a prototype of the energetic, experienced, and uncontrollable salaried manager. He, too, would turn autocrat, albeit somewhat later than Weber, in the mid-1930s, a time when dictators were part of the political environment in Europe. Although McGowan was the undeniable driving force behind ICI, it was Alfred Mond, not him, who carried out the

---

[22] ACS, Emmanuel Janssen, "Voyage aux États-Unis (septembre-octobre 1926)," 28 October 1926.
[23] Chandler, *Scale and Scope*, 358.
[24] Aftalion, *History*, 139.

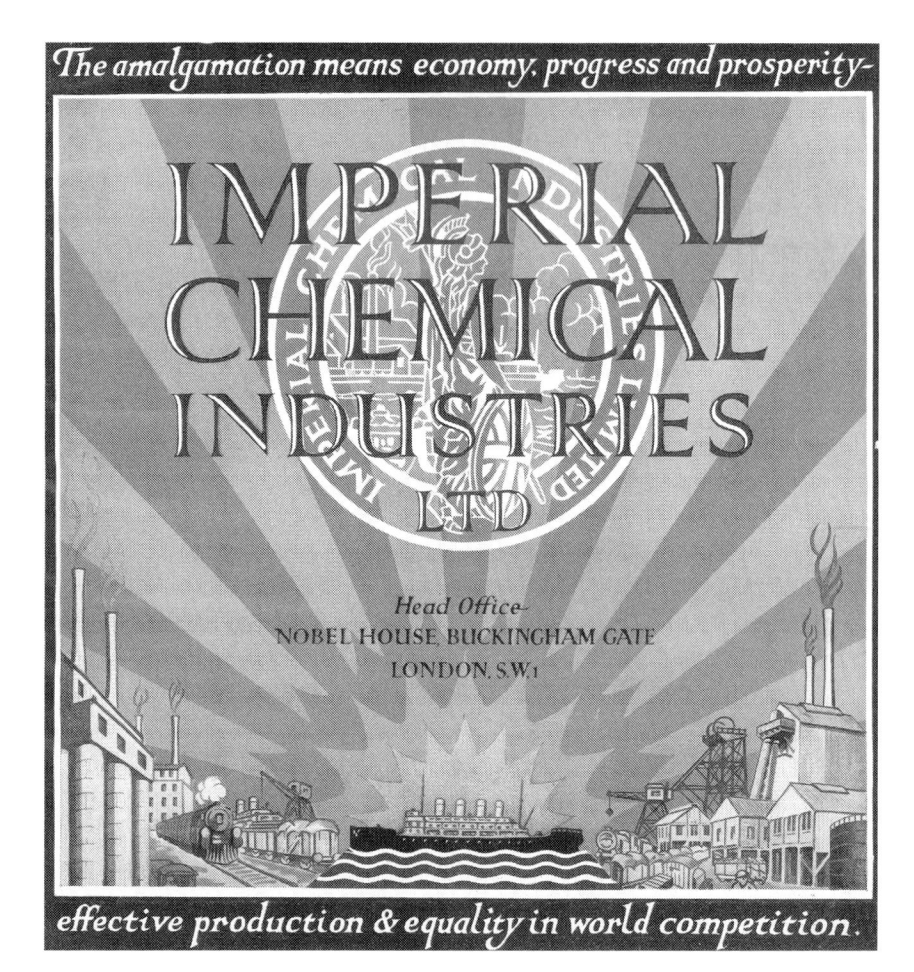

Figure 5.4. Advertisement for Imperial Chemical Industries. Brunner, Mond & Co. was a constitutive part of this British combine, founded in 1926. Through its holding in Brunner, Mond, Solvay & Cie became one of ICI's largest foreign shareholders, with 6 percent of the shares. (ICI archives, courtesy of AkzoNobel)

full reorganization that entailed such an amalgamation. The structure of ICI was based on a blueprint largely designed by Mond, in which centralization and rationalization preceded diversification.[25]

[25] Reader, *Imperial Chemical Industries, Vol. II*, 22–3, 133; Chandler, *Scale and Scope*, 361.

After the transfer of its holding from Brunner, Mond, it appeared that Solvay & Cie held 6 percent of ICI's ordinary stock and was the British combine's largest shareholder. Yet it was not satisfied. Mond had been loyal to the *gérants*, insisting that Brunner, Mond could take on other coalition avenues, should the venture envisioned by Janssen fall through. And that is what happened. As a result, just as six years earlier when Allied was formed, Solvay & Cie witnessed, rather than participated in, the making of a large consolidation. Edouard Hannon could well criticize the fact that the underpinnings of the amalgamation were "more political and nationalist than economic and rational,"[26] it did not alter the essential aspect of the deal: ICI now had the financial capacity to compete at the international level in all kinds of chemical industries (including dyes, at which the British were lagging behind). The "British I.G.," as McGowan called it in reference to I.G. Farben, had an even higher share of the capital than its German counterpart, albeit a much more modest turnover and a smaller workforce.[27] Yet it was in a leading position to engage in the interwar international chemical industry. Even more important to Solvay & Cie, however, was the fact that the ICI partnership also involved a major shift in international alliances, especially with respect to the United States. Because of the existing links between Nobel and Du Pont, it was obvious that the latter would prevail over Allied as far as Anglo-American negotiations were concerned. Overall, Solvay & Cie evidently needed to devise a fresh international strategy.

SPEAKING OF CRISIS: SOLVAY & CIE AT THE END OF THE 1920S

The failure of the international "magic square" holding venture was a hard blow to the *gérants* in general, and to Emmanuel Janssen in particular. Over the years, he had identified with the enterprise. Now that it had collapsed, he was searching for a challenge that could satisfy his appetite for diversification and entrepreneurial partnerships. To some extent, the failed experience rejuvenated his

---

[26] Hannon, "Politique de Solvay & Cie," 8–9.
[27] Cassis, *Big Business*, 36–7.

imagination. He did not start from scratch, however.[28] His preferred instrument was and always had been the Mutuelle Solvay, the family's bank created in 1914 as a public company. Since the end of the war, Janssen had transformed this small-scale private bank into the financial backbone of Solvay & Cie. Through the Mutuelle, he set the stage for the diversification process of the S.A. des Fours à Coke Semet-Solvay. In 1921, he entered in the glassmaking industry with the purchase of the licensing rights for the Libbey-Owens process in Europe. This was a small revolution at Solvay & Cie: For the first time, the company held stakes in soda-consuming industries. It elicited many objections from Solvay & Cie's main consumers in the glass business, especially from the French company Saint-Gobain. But it also aroused the protest of some *gérants* who saw in the Mutuelle's strategy of investment a diversification of Solvay & Cie in disguise. According to them, it contradicted the founding fathers' "alkali-only" principles. Thus, the discussions surrounding the Mutuelle's investment portfolio were insidiously colored with personal issues.

As a matter of fact, the breakdown of the fourfold venture prompted Janssen to launch several initiatives at a time. One was to reorganize Solvay & Cie's interests at Allied into the hands of a "voting trust," Solvay American Investment Company (SAIC), through which he intended to pursue the internationalization of the Mutuelle in view of diversifying the sources of its funding. Investment trusts were very much of the moment. In 1927 alone, 140 investment companies were formed in the United States.[29] The other *gérants*, let alone Weber, had doubts about the initiative, to say the least. But such doubts did not stop Janssen – quite the contrary. The financial investments he conducted through SAIC, in the American glass industry, for instance, had direct repercussions on the Mutuelle's industrial European portfolio. He accelerated the concentration of the Mutuelle's investments in the French chemical and glass industry, which he turned into a kind of holding group in

---

[28] This section draws on Bertrams, Coupain, and Homburg, *Solvay*, chaps. 9 and 10.
[29] John Kenneth Galbraith, *The Great Crash of 1929*, New York, Houghton Mifflin, 1954, 48.

Figure 5.5. Aerial view of the Tavaux plant, soon after its construction in 1932. This soda ash plant, coupled with an electrolytic plant, was built in the French Jura, far from the German border, at the request of the French government, which feared the outbreak of another war. (Solvay Archives)

1927 and proceeded in similar fashion in Belgium the following year. The new company that resulted from the merger of the S.A. des Fours à Coke Semet-Solvay with a series of independent smaller Belgian companies was called Union Chimique Belge (UCB). It was active in all kinds of chemical industries (with the exception of alkali, of course), and 25 percent of its capital was held by the Mutuelle. By 1928, Solvay & Cie's banking instrument had paved the way to diversification of chemical activities *avant la lettre*. On the other hand, a large portion of the Mutuelle, almost half of its interests, was invested in the banking and financial sectors. Industrial positions were depending on transnational financial montages, and vice versa. As long as the international flow of capital was booming, it was considered a virtuous circle of growth generating "great wealth." But it rested on shaky ground.

In December 1928, Solvay & Cie turned sixty-five, the age of retirement. The company surely did not retire but felt the crisis of the retired on its shoulders. The collapse of the American Stock Exchange on 24 October 1929, interrupted the frenzy. This was not the first market crash in history, and nobody knew it would mark the start of a long-lasting social and economic depression that nearly caused the death of capitalism. It did not help Janssen's financial aims, to be sure. The decline and subsequent stagnation of the U.S. market provoked the collapse of the international monetary system and the emergence of a global "credit crunch." This was hardly a surprise: since the end of the war, the Americans had become the world's leading lenders.[30] By 1931, however, the financial crisis turned into a worldwide slump. The Great Depression, as it was called in the United States, only affected Solvay & Cie to a limited extent thanks to its large reserve of assets. But this was not the case for one of the Mutuelle's leading investment banks, which was on the brink of payment default. In-depth reorganization avoided the collapse. The Mutuelle transferred the bulk of its nonchemical investments to a newly created bank and went on a general diet under the close scrutiny of Solvay & Cie.[31]

Important changes also affected the company's top management organization following the death of Armand Solvay on 2 February 1930, after a long illness. Armand had epitomized the union of the family for some time. With his death, there was a void in terms of natural leadership, and Janssen surely did not want to fill it. Since 1928, he had wished to withdraw from Solvay & Cie following strategic divergences with some of his colleagues. He left the company in May 1931 after having waited for the selection of a fine candidate. Following lengthy discussions, René Boël

---

[30] Feinstein, Temin, and Toniolo, *European Economy*, 84–7.
[31] The *Société Belge de Banque* (SBB) was officially created in January 1932 as a result of the divestment of the Mutuelle's holdings in financial and banking sectors and the liquidation of interlocking banking assets. After the major banking reforms of 1934–5, SBB adopted the profile of a corporate bank with direct links with the Solvay, the Janssen, and the Boël families. In 1964, SBB merged with the Banque de la Société Générale de Belgique, which is currently a constituent of the BNP-Paribas-Fortis banking group.

(1899–1990) was appointed *gérant*. He combined all the characteristics a *gérant* could hope for: He was an engineer, had experience in his family's steel business, and, last but not least, was related to the Solvays as Armand's son-in-law. What the *gérants* could not guess, however, was that René Boël's appointment would be one of the best thing that could happen to the company in the years to come.

Figure 6.1. Unemployed workers from the English town of Jarrow during their "Hunger march" to London in October 1936. The financial debacle of 1929 had profound economic, social, and political consequences. (© Illustrated London News Ltd / Mary Evans Picture Library)

6

From Crisis to War

I can see the war that's coming. There are millions of others like me. Ordinary chaps that I meet everywhere, chaps I run into in pubs, bus drivers, and travelling salesmen for hardware firms, have got a feeling that the world's gone wrong. They can feel things cracked and collapsing under their feet.

George Orwell
Coming Up for Air, 1939[1]

There is, in truth, not a formal identity, but a historical bond uniting Stalinism (communism), Nazism (fascism), and New Dealism. Against differing developmental backgrounds and at different stages of growth, they are all managerial ideologies. They all have the same historical direction: away from capitalist society and toward managerial society.

James Burnham[2]

WITNESSING THE AGONY OF CAPITALISM

In the summer and fall of 1931, a sociologist named Wight Bakke decided to study the behavior of unemployed workers in Greenwich, near London. His investigation was based on the methods of participant observation that was popular among sociologists in the 1930s. In Greenwich, Bakke took notice of a young mechanic

---

[1] Quoted in Richard Overy, *The Morbid Age. Britain and the Crisis of Civilization, 1919–1939*, London, Penguin, 2009, 360.
[2] James Burnham, *The Managerial Revolution*, London, Putnam, 1942, 186.

seeking a new job. His dynamism and expertise spoke for him; he was optimistic about his prospects. Three weeks later, the young unemployed worker's optimism began to fail. He had answered newspapers ads, met every condition, and displayed unabashed enthusiasm. Yet all this was fruitless. After eleven weeks, his search for another job was less systematic; doubt won him over. When his job-seeking quest reached three months without success, he felt depressed and negative. The worst was the lack of hope on the horizon, he admitted to the researcher, "It's the hopelessness of every step you take when you go in search of a job you know isn't there."[3]

This unemployed worker from Greenwich was far from the only person expressing his distrust of the economic system as a whole. The Wall Street crash had triggered an unexpected monetary crisis, which was soon transformed into a social and economic slump. By the early 1930s, the industrialized world was engulfed in a huge downward spiral of unprecedented intensity, albeit of varying degrees and differing timing. National governments that had been so prone to intervene in the immediate postwar economic policy were surprised by the shock. During the 1920s, they had desperately sought to return the economic system to normalcy, to the monetary stability embodied by the restoration of the gold standard and laissez-faire politics. These measures proved unsuccessful. Then, as the economic downturn worsened, policy makers adopted a form of economic nationalism through the establishment of high tariffs and the introduction of exchange quotas. Again, such decisions were short-sighted and did not tame the recession but actually accelerated it and increased its impact. And thus, the recession turned into a global depression. Between 1929 and 1935, world industrial output declined by 30 percent, coal and iron production fell by 40 to 60 percent, grain prices dropped by 60 percent, and the volume of intra-European trade exchanges was halved.[4] Only jobless rates were booming. For the first time, the words *unemployed* and *unemployment* were invoked beyond the small circles of sociologists.

---

[3] E. Wight Bakke, *The Unemployed Man. A Social Study*, London, Nisbet, 1935, quoted in Feinstein, Temin, and Toniolo, *European Economy*, 129–30.
[4] Ivan T. Berend, *An Economic History of Twentieth-Century Europe*, Cambridge, Cambridge University Press, 2006, 57–63.

Although available data for the period are not completely reliable, most of the countries exceeded 20 percent unemployment among the workforce.[5] In Germany and the United States, one worker in three was registered as unemployed in 1932–3, the peak years of the crisis. Not surprisingly, both Adolf Hitler and Franklin Delano Roosevelt made employment their top priority when they each came to power in January 1933. This was an innovation in many regards. The social system was the last domain where the politics of laissez-faire had prevailed.[6]

Whereas the magnitude of the depression was particularly severe in traditional industrial sectors (coal, metallurgy, textile), the chemical industry proved relatively more resilient. As noted in the previous chapter, the financial downturn directly affected Solvay & Cie's investment portfolio and prompted the company to reorganize its banking activities, but from an industrial standpoint, the effects of the depression were milder. This is not to say that Solvay & Cie was immune to the crisis. Between 1929 and 1933, alkali sales dropped by an annual average of 20 percent. In terms of soda ash output, factories well situated for overseas export outdid those located in the traditional industrial nexus; for instance, the Torrelavega and Rosignano factories, in Spain and Italy, respectively, displayed better resistance to the depression than Belgian and French factories. As it happened, the factories at Dombasle, Giraud, and Sarralbe did not reach their production levels of 1929–30 until after the Second World War. Thus, alternatives were sought and opportunities exploited.

Shortly before Dombasle and the other French plants were hit by the crisis, the *gérants* had decided to erect a new plant at Tavaux, in the French Jura. The French government had strongly lobbied for the location because it helped balance the geographic layout of the chemical industry in France. The Tavaux factory was to become the cornerstone of the Solvay & Cie's industrial strategy in France, focusing on electrolytic caustic soda. In many ways, the plant was a bet on the future as well as a confirmation of decisions made in

---

[5] Charles H. Feinstein, Peter Temin, and Gianni Toniolo, *The World Economy between the World Wars*, Oxford, Oxford University Press, 2008, 124–5.
[6] Kiran Klaus Patel, *Soldiers of Labor. Labor Service in Nazi Germany and New Deal America, 1933–1945*, New York, Cambridge University Press, 2005, 32–9; Richard J. Evans, *The Coming of the Third Reich*, London, Penguin, 2003, 236.

the past. Solvay & Cie had purchased the Castner-Kellner mercury-cell process back in 1895 and had built several electrolytic units in the meantime (the Jemeppe, Lissitchansk, and Torda plants).[7] But the unit that was on stream at Tavaux in September 1930 marked a renewed impetus in the electrolytic domain. It not only strengthened the company's strategy in the production of caustic soda, it also took up the challenge of producing chlorine at a time when the demand for chlorine-based products was booming.[8]

By and large, the blow of the depression was less painful at Solvay & Cie than it was for its partners I.G. Farben or ICI. At BASF, for instance, the total employment of Ludwigshafen and Oppau factories fell from 24,000 in 1924 to a low point of 12,300 in 1932.[9] Despite the lack of evidence, it can be argued that Solvay & Cie did not experience a shrinkage of its workforce to a similar degree. Where it was possible, a reduction of working hours was introduced as a means to offset overproduction and lagging demand. But here again, practical situations differed largely from plant to plant, in view of national conditions. Numerous strikes took place during the interwar period (with peak years in 1919 and 1920 and general instability throughout the 1930s),[10] and serious social turmoil was brewing. Providing workers new social-security perspectives thus became a priority for industrialists after the First World War. For Solvay & Cie, the challenge above all lay in adjusting its tradition of social benefits to the context of growing nationalism.

RECASTING INDUSTRIAL STABILITY

Solvay & Cie had not waited for the postwar years to introduce welfare benefits for its workers and employees. Since the 1880s, the rationale of the company's social policy was embedded in the principles of paternalism that were quite in line with the zeitgeist of corporate-based social measures. It consisted in providing workers with various social welfare incentives in addition to their wages (limited health insurance, retirement schemes, and housing, for

---

[7] Bertrams, Coupain, and Homburg, *Solvay*, chap. 5.
[8] Bertrams, Coupain, and Homburg, *Solvay*, chap. 11.
[9] Abelshsauser et al., *BASF: the History of a Company*, 245.
[10] Beverly J. Silver, *Forces of Labor. Workers' Movements and Globalization since 1870*, New York, Cambridge University Press, 2003, 125–6, 143–4.

example). In return, labor was expected to maintain the social peace. In the eyes of the management, the underlying objectives of this approach were to stabilize and bring discipline to the workforce, which had demonstrated high rates of turnover. At Solvay Process Company's Syracuse plant, a third of the workers in 1919 had been active at the factory for fewer than three years. At the Detroit plant, this group comprised 56.35 percent of the personnel. Half were recent immigrants who attended the company-located "schools of Americanization" to learn English and American "values."[11]

Thus, the management sought to reinforce the loyalty of labor at the workplace. As one historian noted, "When workers heard that another company had entered a seasonal rush and was hiring at better pay, they quit the jobs they had. If rumors circulated that an employer would soon be laying off, they did not wait to get the official news. When a worker became angry at his foreman, he left if he had anywhere else to go."[12] As a result, negotiations among local employers enabled management to avoid outbursts of social competition. The introduction of new welfare initiatives frequently arose from such negotiations. In this respect, the company took into account the sometimes very different social patterns that existed on a national or regional level and transferred them into its plants. In an effort toward decentralization, plant managers played a crucial role in this respect. For instance, it was the Donetz plant manager, Sigismond Toeplitz, who took the initiative to introduce the three-shift, eight-hour day system in the factory in 1897. Following the promising results in increased productivity, this policy was later generalized to the Solvay group as a whole.[13]

World War I ushered in a period of State-based social reforms. Many of these had been anticipated and implemented earlier by Solvay & Cie. Like other industrialists, however, the *gérants* observed the State-driven ambition to homogenize and nationalize the fragmented environment of social policy once dominated by private initiatives. Highly debated issues focused on the questions of industrial relations, collective bargaining agreements, and State

---

[11] ACS 1271–33-1 A, SPC Report of Activities, 12 September 1919.
[12] Lizabeth Cohen, *Making a New Deal. Industrial Workers in Chicago, 1919–1939*, New York, Cambridge University Press, 1990, 197.
[13] Bertrams, Coupain, and Homburg, *Solvay*, chap. 4.

arbitration. Labor-management cooperation was the keyword of these discussions. Solvay & Cie viewed such cooperation positively inasmuch as it was initiated directly by workers' representatives rather than by outside organized labor unions. In 1919, at Dombasle, the plant manager agreed to support "a trade union exclusively composed of workers belonging to the Solvay plants of the region" only to regret later that such "autonomy [did] not exist" at the time.[14] In Germany, DSW had to cope with the passing of the Factory Council Law in February 1920, which institutionalized labor committees modeled on workers' and soldiers' councils in Russia, the so-called soviets. Of course, "the acceptance and effectiveness of factory councils varied from industry to industry, from firm to firm, from director to director, and from time to time."[15] In that context, the management of DSW did its best to adapt the State-based social engineering to the company's own welfare tradition. Hence, it expanded the framework of the Ernest and Alfred Solvay Fund, created in the 1890s as a retirement scheme for workers and employees. In an era increasingly filled with political intensity, it appeared that Solvay & Cie's top and middle management intended, on the contrary, to *depoliticize* labor issues. In the end, however, this proved pointless, especially in countries where dictatorial regimes came to power.

ITALY AND GERMANY: LABORATORIES OF FASCISM

From the early 1930s onward, the principles of paternalism were superseded by those of corporatism. Whereas paternalism was characterized by a bilateral labor-management relationship, corporatism was a system of bureaucratic multilateralism orchestrated by the State. It entailed a relative loss of power for companies in crafting social measures. Solvay & Cie management, in other words, could no longer set the terms for workers inside the business but had to negotiate and coordinate with a dense web of external agencies. The State-driven corporatist organization gained momentum in Italy under Benito Mussolini. The Fascist regime had

---

[14] ACS, minutes of the meetings of the *gérance*, 4 April 1919.
[15] Fear, *Organizing Control*, 426.

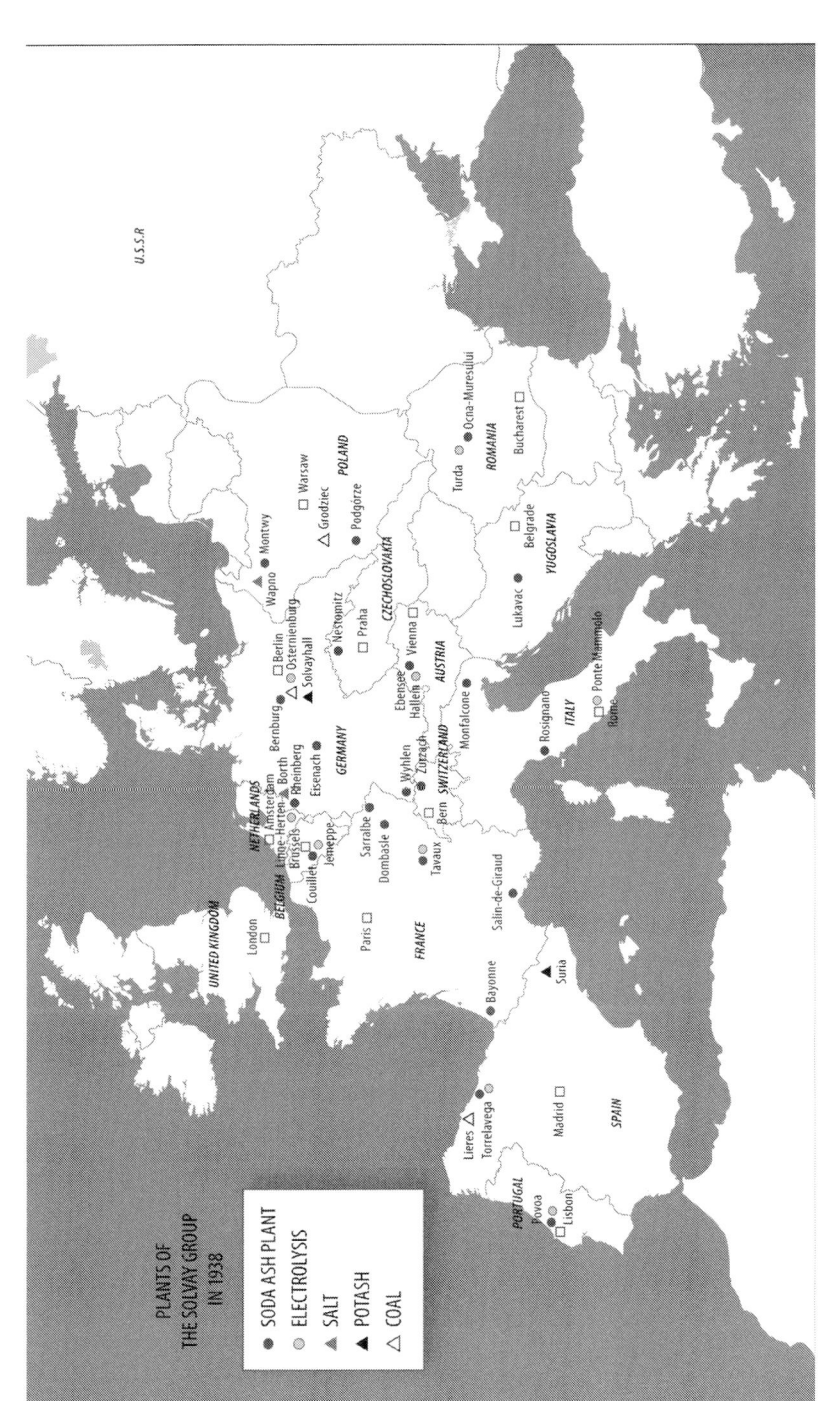

Figure 6.2. Map of the Solvay plants in 1938.

promulgated a Labor Charter as early as April 1927, yet it was not until the mid-1930s that the corporatist structure was fully implemented in regulating labor policy and the Italian economy as a whole. In Italy, Solvay & Cie relied on delegates within the Corporation of Chemical Industry, one of the twenty-two corporations created by the National Council of Corporations. Through the corporatist scheme, Mussolini not only intended to find a way out of the crisis – the *grande depressione* that hit the Italian industry hard – he also sought to establish a "third way" between communism and capitalism, a new economic architecture that would be compatible with State interventionism, self-sufficiency, and (some) private initiatives.[16]

The Italian Fascist blueprint brought State interventionism to unprecedented levels, but this was an illustration of a wider trend visible everywhere in Europe and, to some extent, in the United States as well.[17] Next to the founding of State-owned companies, the introduction of State planning paved the way to the nationalization of industrial and financial sectors affected by the crisis. Influenced by the Russian five-year plan as well as by the American ideology of technocracy and scientific management, the Belgian socialist leader, Hendrik De Man, vigorously campaigned for the adoption of the Labor Plan at Belgium's general election of 1933.[18]

In Brussels, the *gérants* were wary of the growing manifestations of State intervention in Europe, which they associated with a form of dirigisme worthy of Colbert, the economic minister of Louis XIV of France. Determined to reorganize Solvay & Cie's social policy, Louis Solvay had in early 1932 created a kind of think tank, the Maison Ernest Solvay, which acted as a steering committee regarding the company's labor and welfare measures. The members of the "brain trust" gathered by Louis Solvay wished to learn more

---

[16] Alessio Gagliardi, *Il corporativismo fascista*, Roma, Laterza, 2010, 70–3; Alberto De Bernardi, *Una dittatura moderna. Il fascismo come problema storico*, Milano, Mondadori, 2006, 179–81.

[17] Wolfgang Schivelbusch, *Three New Deals. Reflections on Roosevelt's America, Mussolini's Italy, and Hitler's Germany, 1933–1939*, New York, Picador, 2006.

[18] Zeev Sternhell, *Ni gauche, ni droite. L'idéologie fasciste en France*, Bruxelles, Complexe, 1987, 331–57; Mario Telò, *Le New Deal européen. La pensée et la politique sociales-démocrates face à la crise des années trente*, Bruxelles, Editions de l'ULB, 1988, 101–10; Vinen, *History in Fragments*, 165–7.

about the Fascist organization of the economy. As a result, they asked the plant manager of Rosignano, Clément Van Caubergh, for more details. Corporatism, Van Caubergh told them, was certainly not intended to "defend the interests of industry"; it was a way to ensure the peaceful collaboration of labor and management. "The Italian totalitarian method," he went on, "rests on the principle that individual interest should be subordinated to general interest." Although Van Caubergh believed the corporatist organization was largely beneficial to a company like Solvay & Cie, he did not recommend its transfer to other "political climes." In fact, the corporatist organization "can only grow usefully where there is no rivalry between political parties, no compromise concluded, and where a strong government feels itself capable of enforcing its laws, which give the official trade unions the power and the tranquility they required."[19]

Far-reaching as it was, the Fascist model of interventionism was a dwarf compared with its German counterpart.[20] In fact, in Hitler's regime, it was State intrusion, not intervention, that was most evident. Reorganization of the economy under the Nazi regime was so sweeping that it derived from a "profound organic transformation," as DSW head manager Ernst Eilsberger put it in a report he sent to the Maison Ernest Solvay in 1935. Among various observations, Eilsberger underlined the immediate suppression of all the trade unions and workers' associations that existed before 1933 and their replacement by a single body, the German Labor Front, the only structure that was tolerated. The Nazi regime also endorsed "works communities," the loyalty of which was undisputed, contrary to the labor committees that had been created during the early days of the Weimar Republic. These communities were themselves run by the "plant leader" (*Führer des Betriebes*) in a hierarchical manner. Nine *Betriebsführers* were designated at DSW, and they represented all units of operation. Although DSW management still had the upper hand when it came to control of the personnel, there was no doubt that everyone and everything had

---

[19] For this and other quotes, see Bertrams, Coupain, and Homburg, *Solvay*, chap. 12.
[20] Adam Tooze, *The Wages of Destruction. The Making and Breaking of the Nazi Economy*, London, Penguin, 2006, 37–49.

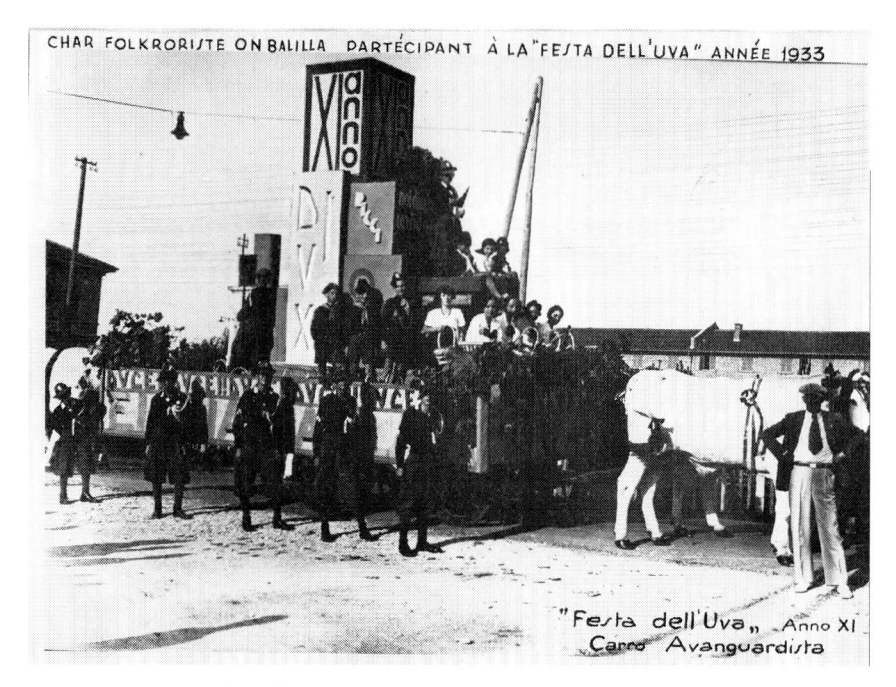

Figure 6.3. A folk float decorated with fascist symbols, during the Festa dell'Uva (Grape Festival) in the Rosignano-Solvay village (1933). (Solvay Archives)

to be in line with the military outlook shaped by Nazi legislation. Among the variety of prescripts issued by the local German Labor Front authority, responsible for the Bernburg plant, one could read the following: "Promote to all productive men the clearest conception on the nature and goal of our universal revolutionary aim, i.e., increase the joy of living and working, transforming the idle and dull worker at the factory into a honest, free, and joyful German worker; these are the tasks of our plants groups." Not surprisingly, the intrusion of the Nazi State went well beyond the border of the workplace. It made a regime priority of controlling working-class leisure. Mussolini had once again pioneered in this area with the introduction of official "leisure associations" (known as the *Dopolavoro*) in 1925. The Nazi leisure program, established in 1933, was known as "Strength through Joy" (*Kraft durch Freude*), the ironic wording of which is reminiscent of the infamous motto

Arbeit macht frei ("work brings freedom") that was engraved on the iron gates of the Auschwitz concentration camp.[21]

These measures actually found a concrete application at Solvay & Cie's factories in Italy and Germany, where plant managers strove to legitimize the ongoing reforms. At DSW, for instance, Eilsberger took pains to ensure a firm continuation between the "Solvay tradition" of social-welfare initiatives and the principles of Hitler's New Order. The problem was, as any well-informed person knew, that DSW was overwhelmingly controlled by non-German capital. Consequently, Eilsberger had to explain to the workers that, despite its Belgian shareholders, DSW was run by "German men" and conducted "according to the German spirit." Workers, employees, and managers active at DSW "can thus proudly raise their head as *Germans* and refute with full knowledge of facts the shortsighted views which pretend to [portray] us.... [as] a foreign company." To what extent such rhetoric convinced the workforce is difficult to tell. Any expression of discontent was forbidden and violently repressed. Reporting on the social and economic organization of Nazi Germany, a delegate from the Committee of Belgian Industrialists summed up the contradiction: there was "an indisputable atmosphere of enthusiasm at the activities organized by the German Labor Front," he wrote in October 1936. On the other hand, he went on, "These results were not achieved without overcoming difficulties; it sometimes required the adoption of very serious disciplinary measures (discharge of business managers, labor suppression, transfer into concentration camps)." This was still a moderate observation. Shortly thereafter, Hitler's approach to social policy became more racially than socially oriented. The biological enemies to the German "national community" were to be suppressed, period.[22]

1936: THE SPANISH PRELUDE

With the benefit of hindsight, it appears that the bloody civil war that erupted in Spain in the summer of 1936 was a kind of general

---

[21] Vinen, *History in Fragments*, 169–70.
[22] Mazower, *Dark Continent*, 99–102.

rehearsal to the Second World War. The Spanish Civil War epitomized the political tensions and social turmoil that shook Europe starting in the early 1930s. It represented the event on which the relics of the postwar international order backfired – a universal symbol of the political drift that separated the world into two irreconcilable camps. "Never since the French Revolution," wrote a British poet, "had there been a foreign question that so divided intelligent British opinion."[23] Advocates of moral order, stability, and the Church approved General Francisco Franco's Nationalist cause, which grew stronger after a strategic, albeit failed, coup launched on 17 July 1936. On the other side, there was a broad antifascist movement to support the official republican government. Outside Spain, people from different backgrounds would energetically throw themselves into a campaign in support of the republic. This was done by sending volunteers, the well-known International Brigade, to the battlefield or through politically committed literature, the kind of which Ernest Hemingway, George Orwell, and André Malraux provided.

Spain's political instability was not unknown to Solvay & Cie's *gérants*. Since 1930, which coincided with the end of a military dictatorship, riots and protests took place at the Torrelavega plant as at other workplaces in the northern region of the country. The most serious turmoil occurred in October 1934 in the wake of the Asturian miners' "Revolución." Torrelavega was the stage of a violent clash between local authorities and anarchists, which resulted in the dismissal of 221 "agitators." It represented no less than 25 percent of the factory's workforce.[24] Torrelavega, as a result, undertook a "political cleansing" well before the outbreak of the civil war. When Franco's troops occupied the region in August 1937, no one really opposed them, and the plant promptly resumed its activity under the close scrutiny of nationalist rebels. This was not the case at Suria, Solvay & Cie's other plant in Spain, located in Catalonia. The potash mines of Suria had been purchased for strategic reasons in 1919 after lengthy negotiations with French and American

---

[23] Quoted in Overy, *Morbid Age*, 320.
[24] Ángel Toca, *La introducción de la gran industria química en España. Solvay y su planta de Torrelavega (1887–1935)*, Santander, Universidad de Cantabria, 2005, 258–60.

Figure 6.4. Workers digging potash in the Suria mine in Barcelona's hinterland. On the surface, the Spanish Civil War took place between 1936 and 1939. Suria was in the Republican zone, whereas the Torrelavega soda ash plant was in the Nationalist zone. (Solvay Archives)

competitors. Production started only in October 1921 following major investments. The least one could say was that it was not the company's most profitable venture. As soon as the civil war erupted, workers at Suria showed increasing signs of insurrections. The Belgian plant manager, Norbert Fonthier, had to be evacuated together with several members of the personnel.[25]

In view of the magnitude of the Suria incident, the issue took a political turn. Stressed by the threat of nationalization, the *gérants* initiated protest through official diplomatic channels. At the time, the Belgian government was committed to a regime of political neutrality in terms of foreign affairs, that is, to noninterventionism when it came to Spain. Under the pressure of Belgian industrialists with important interests in Spain, the government's attitude progressively shifted toward diplomatic recognition of Franco's

---
[25] See Bertrams, Coupain, and Homburg, *Solvay*, chap. 11.

regime. This was concluded in November 1938 after an overheated speech by prime minister and minister of foreign affairs Paul-Henri Spaak before the Belgian Senate. In January 1939, Franco's troops marched on the region of Barcelona. This contributed to ending the war. The Suria plant began operation after a long interruption, and both Solvay & Cie's plants averted nationalization. Yet, as Solvay & Cie's management would soon find out, the issue of sequestration – that is, confiscation through military custody – was to be regularly addressed in the months to come.

HITLER'S GREATER GERMANY

On 11 March 1938, the Austrian chancellor Kurt von Schuschnigg was forced out and replaced by the interior minister, a prominent pro-Nazi official. For months, Adolf Hitler had campaigned to win the hearts and minds of his "fellow" Austrians to set the stage of Austria's annexation by Germany, the so-called Anschluss. In Vienna and elsewhere in the country, Nazi propaganda banners carried quotations from the Führer's writings – "Those of the Same Blood Belong in the Same Reich!"[26] Solvay & Cie's management intended to react quickly, and the general direction of the Solvay-Verein combine in Central Europe was immediately transferred from Vienna to Prague. The relief was short-lived, however. The Anschluss was the first step toward Hitler's total hegemony over Europe. In October 1938, German troops marched in the region of the Sudetenland in Czechoslovakia, where the Nestomitz Solvay-Werke plant was located. All of a sudden, the powerful chemical concern I.G. Farben filed an authorization to acquire the Verein Aussig's directly owned plants. The Verein showed unexpected resistance but eventually had to transfer the control of its factories to a newly created German company. I.G. Farben thus lost the deal but promised to try again, later.[27] The political climate for foreign businesses was at its worst. Just as they had done for the Spanish case some months before, the *gérants* expressed their worries to

---

[26] Quoted in Mark Mazower, *Hitler's Empire. Nazi Rule in Occupied Europe*, London, Penguin, 2008, 51.
[27] Peter Hayes, *Industry and Ideology. I.G. Farben in the Nazi Era*, New York, Cambridge University Press, 1987, 238–9.

Belgian diplomats. After all, Solvay & Cie held 15.4 percent in Verein, which made Belgium the largest country in terms of foreign direct investment in Czechoslovakia's chemical industry.[28] At that time, however, the international community was still blinded by Hitler's promises to unite German minorities abroad in a peaceful way. Thus, when Hitler offered to "protect" Czechoslovakia as a whole in March 1939, he heard no opposition from Western European countries. The Czechs were left alone.

The advance of the Nazi power had also direct repercussions at DSW. Since 1937, two men close to the regime, Rüdiger Graf von der Goltz and Carl Adolf Clemm, had made their way up in the company's hierarchy. By January 1939, the latter had been designated chairman of DSW's executive board (*Vorstand*), and the former had been appointed member of its supervisory board (*Aufischtsrat*), which had been "Germanized" in the meantime. The Germanization process of boardrooms was complementary to the biologically-driven Nazi laws of Aryanization, which sought to eliminate "non-Aryans" from public, and then from private positions. Since 1938, the pressure of Aryanization increased immensely, focusing specially on the eviction of Jews from business activities. The Czech delegate of the Verein at Ebensee plant, Victor Basch, was to be the victim of this official discrimination, and he was surely but one among many. The intensification of the exclusion policy prompted the *gérants* to send their key man in the region, Eudore Lefèvre, to Zürich, where he managed to run Solvay's interests in southeast Europe during the war. On the other hand, they entrusted DSW head manager Adolf Clemm extended powers to ensure the control of the company's interests in Eastern Europe. They thought that Clemm's action would be limited to the factories located in Austria and Czechoslovakia. They were wrong: Hitler's army invaded Poland on 1 September 1939. Sixteen days later, conforming with the secret signing of the Molotov-Ribbentrop Pact, the Red Army entered the Polish territory from the eastern side.

The factories of Zaklady Solvay in Poland were thus located in what historian Tim Snyder has coined the "Bloodlands," the

---

[28] Alice Teichova, *An Economic Background to Munich. International Business and Czechoslovakia, 1918–1938*, Cambridge, Cambridge University Press, 1974, 279–81.

Figure 6.5. After the Anschluss of Austria by the Germans in 1938, Vienna was under Nazi rule. Geopolitical reconfigurations of Europe had a long-lasting impact on the Solvay Group. (Mary Evans Picture Library/Robert Hunt Collection)

territory of Central Eastern Europe crushed by Hitler's and Stalin's respective mass killings. Both the soda plants of Mątwy (south of Inowrocław) and Podgórze were west of the Molotov-Ribbentrop line, where "Jews came under German control significantly earlier, but were killed later."[29] In the first weeks of the Nazi attack, the occupation of Inowrocław involved arrests, expulsions, and killings of many local inhabitants, but the worst came after June 1941, which coincided with the full invasion of Poland by Hitler's troops. From December 1942 to August 1943, the Mątwy soda plant was converted into a Jewish forced labor camp that accounted for thousands of casualties. In Galicia, south of the city of Krakow, the Podgórze Solvay factory processed limestone, which came from a quarry that used forced manpower from the nearby Plsazow labor camp. In 1993, the quarry in question served as film set, and the camp as a reference model, for the shooting of Steven Spielberg's acclaimed movie, *Schindler's List*.[30] Off the "Bloodlands," dreadful wartime spots would also overlap with Solvay's own history. These included the Ebensee concentration camp in Austria, near the soda plant jointly owned by Solvay and the Verein Aussig until the Anschluss. At Ebensee camp, 27,000 inmates were imprisoned, a third of which died due to inhuman working conditions.[31]

THE ECONOMICS OF OCCUPATION

When Hitler's troops invaded Belgium in May 1940, at the same time they were invading Luxembourg and the Netherlands, the population had a feeling of déjà vu. "Twice in one generation is pretty stiff," wrote the British politician Hugh Dalton; "now again, as twenty-five years before, German hands pulled the levers that [have] launched the Death Ship."[32] Yet 1914 did not have a lot in common with 1939–40. With regard to the occupation of Belgium,

---

[29] Timothy Snyder, *Bloodlands. Europe between Hitler and Staline*, New York, Basic Books, 2010, 117–18, 253.
[30] I owe this information to Wikipedia pages.
[31] See Florian Freund, *Die Toten von Ebensee. Analyse und Dokumentation der im KZ Ebensee umgekommenen Häftlinge 1943–1945*, Dokumentationsarchiv d. Österr. Widerstandes, Vienna, 2010.
[32] Quoted in Overy, *Morbid Age*, 314.

there is no doubt that differences largely outweigh similarities. First, Germany's warfare experience enabled the occupying authorities to avoid past mistakes. A basic principle, which derived from Hitler's military strategy of *Blitzkrieg*, was to foster collaboration with the local population and encourage partnership with the national economy.[33] Second, the large-scale humanitarian program that had been launched by Solvay and Heineman and organized by Francqui and Hoover during the First World War – the National Committee for Relief and Food and its U.S. equivalent the Committee for Relief in Belgium (see Chapter 4) – could not be reproduced to the same degree. Winston Churchill was staunchly determined to maintain a tight blockade of the continent. This and other considerations crossed Ernest-John Solvay's mind as he was mulling over the current situation with the main economic actor of the country, Alexandre Galopin. Belgian industrialists were somewhat in the dark as to how to proceed during wartime. In their rush toward London, the members of the government had given ambiguous instructions. As head of the powerful Société Générale de Belgique, Galopin justified the resumption of work as a means to avoid starvation of the population and forced labor.[34] Ernest-John Solvay intended to follow this line of action. It might have reduced the number of forced laborers, but it did not prevent the organization of forced labor in general, for instance, at the Bernburg, Rheinberg, or Hallein plants.

Dealing with the occupying authorities was one thing; having a clear say in business decisions was another. Ernest-John Solvay was rapidly informed that DSW head manager, Adolf Clemm, had been promoted "administrator" (*Verwalter*) of Solvay & Cie's interests in the rapidly increasing German-occupied territory. Moreover, Helmut Eilsberger, the son of the current chairman of DSW supervisory board, Ernst Eilsberger, had been given a special mandate for the production of soda on the French territory. The German "supervision" of Solvay & Cie's interests thus lay in the hands of DSW-related higher management. But to what extent were people

---

[33] Alan S. Milward, *War, Economy, and Society, 1939–1945*, Berkeley, University of California Press, 1977, 132–3.
[34] Herman Van der Wee and Monique Verbreyt, *A Small Nation in the Turmoil of the Second World War*, Leuven, Leuven U.P., 2009, 100–3; Patrick Nefors, *La collaboration industrielle en Belgique, 1940–1945*, Bruxelles, Racine, 2006, 5–10, 29–38.

Figure 6.6. Ernest-John Solvay and René Boël, the two Solvay managers in charge of relations with the authorities in occupied Europe and the free world, respectively, during World War II. (Solvay Archives)

like Clemm, the younger Eilsberger, and von der Goltz really helpful, and loyal, to the company? This, Ernest-John Solvay soon would learn when an old acquaintance of Solvay & Cie rapidly showed interest in absorbing its subsidiaries and interests abroad. The German chemical combine I.G. Farben, as we know, was interested in Solvay's investments in southeast Europe, which it partly controlled with the Verein. Now that the Verein had been silenced and Solvay was "administered" by Germans, I.G. Farben felt the moment was appropriate. Against all odds, I.G. failed again. Von der Goltz and Clemm had mobilized their political networks and persuaded them to prevent the giant trust from becoming an objective threat to other German chemical companies. Additionally, behind Goltz and Clemm, a third and somewhat controversial character had been pulling useful strings: the managing director of Deutsche Bank and recent member of DSW's supervisory board, Hermann Josef Abs. For economic and strategic reasons, Abs had recommended prudence and sobriety when it came to facilitating

German capital penetration in Western Europe.[35] Whatever the underpinnings of the negotiations, the campaign largely paid off: DSW, and Solvay & Cie's interests as a whole, were protected by a form of "hands-off policy" guaranteed by the Reich's Ministry or Economic Affairs.[36] The same fight for autonomy, and against nationalization, was conducted in Italy by Ernest-John Solvay. With the help of some competitors, the chemical company Montecatini attempted to overthrow Solvay & Cie's interests there.[37] The operation eventually backfired, notably because of Ernest-John Solvay's far-reaching odyssey from the vortex of Fascist bureaucracy up to Mussolini himself. The Montecatini episode had revealed the harmful overlap and confusion of interests that took place between business and politics in dictatorships.

MEANWHILE, ACROSS THE OCEAN . . .

René Boël was not aware of the *Verwaltung* system, nor of the hostile takeover attempts Solvay & Cie was facing. In fact, he did not spend the war with his colleagues in occupied Belgium. He devoted the bulk of his time as "*gérant*-in-exile" reorganizing Solvay's networks overseas and organizing them along the lines set by the Allies. As he arrived in England in June 1940, however, Boël's first action had not been to focus on Solvay & Cie. Working under the direction of the Belgian minister of finance (in exile), Camille Gutt, he was in charge of supervising various trade and industrial agreements between the Allies and occupied Belgium. Yet Solvay & Cie was also threatened in Great Britain. The Belgian company was considered enemy property. With the help of their friends at ICI, Boël managed to avoid allowing Solvay's capital in Great Britain to fall under the restrictions of the "Trading with the Enemy Act." What Boël feared the most, however, was not the British situation; it was the American attitude. During the First World War, the American "Allies" had seized the opportunity of a "European conflict" to assert their independence. With Orlando Weber still active behind the

---

[35] Tooze, *Wages of Destruction*, 390–1.
[36] Hayes, *Industry and Ideology*, 276.
[37] See Bertrams, Coupain, and Homburg, *Solvay*, chap. 12.

scenes at Allied, the worst was always an option. Back in 1934–5, Boël had teamed with two experienced legal advisers, John Foster Dulles and George Murnane, to oust the uncontrollable manager. The maneuver had been painstaking but proved successful in the long run: Weber was forced to resign as CEO. Nevertheless, the psychological warfare between both men gave way to strong feelings of resentment. Shortly before the invasion of Belgium, in April 1940, Weber had led a putsch at the board of Allied denouncing the "un-American" behavior of Solvay's representatives. Boël was urged to travel to New York and solve matters directly, which he did. Now that Belgium was defeated, Boël expected the outbreak of another coup. Hence, he decided to leave London and pay his American partners a friendly visit in October 1940.

Contrary to his expectations, however, there was no sign of a Weberian plot. Instead, the combination Boël had set up in April 1940 was still firmly established. On the other hand, the nationalist atmosphere was overwhelming; it prompted Foster Dulles to advise the *gérant* to keep a low profile during his stay in the United States. Boël's activity was thus divided between two contradictory poles: acting below the radar on the one hand, and striving for legal recognition of his position as *gérant* of Solvay in territories not under German control on the other. When he finally achieved both these goals, which took an inestimable amount of time, he started to exert direct managerial control on Solvay's plants located in neutral countries in Europe (Spain, Portugal, Switzerland). His second objective was forward-looking. In partnership with ICI, Boël led Solvay & Cie to launch a venture in Brazil, a country that enjoyed considerable growth at the time. The deal, however shaky during the mid-1940s, paved the way for a large-scale redeployment of Solvay's position in the postwar international trade of alkali and chlorine products.

Overall, Solvay & Cie's two wartime experiences produced unparalleled and quite opposite views: the view from German-occupied Europe versus the view from Anglo-America. Bridging these views was an uneasy task. Shortly after the First World War, Solvay had discovered the emergence of a new world. Likewise, the second postwar would bring the company plenty of unexpected challenges.

Figure 7.1. Solvay was the first company to produce and commercialize plastic bottles for mineral water. (Solvay Archives)

7

Reconstruction through Diversification

This war is not as in the past. Whoever occupies a territory also imposes on it his own social system. Everyone imposes his own social system as far as his army can reach. It cannot be otherwise.

Joseph Stalin[1]

– Vice-President Nixon: *We don't have one decision made at the top by one government office.... We have many different manufacturers and many different kinds of washing machines so that the housewives have a choice.... Would it not be better to compete in the relative merits of washing machines than in the strength of rockets?*
– First Secretary Khrushchev: *Yes, but your generals say: "We want to compete in rockets." We can beat you.*
Nixon-Khrushchev "Kitchen Debate"
Moscow Trade Fair, June 1959[2]

THE RESETTLEMENT OF POSTWAR GERMANY

The liberation of Europe by the Allied armies marked the end of one of the darkest pages of the world's history. After having reached a territory that stretched from Bayonne to Leningrad and from Copenhagen to the Donets River, Hitler's empire collapsed from the summer of 1944 onward under the irrepressible advance of the Anglo-American troops in the West and the Red Army in the East.

---

[1] Quoted in Mazower, *Dark Continent*, 217.
[2] Quoted in Victoria de Grazia, *Irresistible Empire. America's Advance through 20th-Century Europe*, Cambridge, Belknap Press, 2005, 456.

In April 1945, the National-Socialist hegemony over Europe was definitively terminated. Yet transforming Europe into a Nazi-free zone did not entail that peace and democracy were brought back to the continent in the blink of an eye. For many reasons indeed, 1945 was a dreadful year. In northern France, Belgium, and the Netherlands, Allied armies promptly realized that local populations were starving. "Everyone was worn out; many died along the road," reported a Dutch girl in the early spring of 1945. "The farmers would not give anything for money and therefore the people gave their last shoes, their last coat, for just a little bit of food."[3] The situation was even worse in Central and Eastern Europe. Reaching Poland, Red Army soldiers discovered massive death camps. These were not the "usual" concentration camps implemented by the Nazi regime since the 1930s and reproduced almost everywhere in Europe as a means to isolate target groups and condemn them to forced labor. These were death factories – the sites where Hitler and his crew had ultimately decided to eliminate Jews from Europe.[4]

At the same time, however, all eyes converged on Germany. As anti-German hatred and desire for revenge ranked high among recently liberated populations, Allied political authorities were negotiating the future of the country. One thing was clear: Everyone sought to avoid the mistake of 1919, that is, the planned annihilation of the defeated belligerent. At the Yalta Conference in February 1945, the three great Allied Powers endorsed the establishment of four provisional occupation zones – one for each power and one for France – in Germany and Austria, as well as in Berlin and Vienna. Roosevelt and Churchill gave in Stalin's demand to reduce the size of the prewar Reich by 25 percent and to obtain reparations from Germany. Contrary to a widespread mythology, Yalta did not legitimate the partitioning of Europe into two opposite blocs. It did, however, illustrate Stalin's frightening appetite over Europe and the reluctance of Anglo-Americans to counter his ambitions.[5] A second

---

[3] William I. Hitchcock, *Liberation: The Bitter Road to Freedom, Europe 1944–1945*, New York, Free Press, 2008, 111.
[4] Snyder, *Bloodlands*, 253–76.
[5] William I. Hitchcock, *The Struggle for Europe: The History of the Continent since 1945*, New York, Doubleday, 2003, 19–23.

meeting was planned to solve the "German issue." The Big Three decided to convene at Potsdam, in the outskirts of Berlin, in July 1945. All representatives firmly agreed to keep Germany as a single unit, yet they all demanded full control and power in their respective zones of occupation. This was not the last contradictory legacy of Potsdam. Another dazzling ambiguity dealt with the practical meaning of the "four D's" to which Germany had to submit: demilitarization, democratization, denazification, and decartelization. The problem was that each occupying authority had a different, if not divergent, view on how to apply these concepts. Overall, Potsdam delivered an incoherent outline. It did not take long for actors to perceive this.

As was the case during wartime, DSW turned out to be a laboratory of European history.[6] And just like after World War I, the fate of DSW during the second postwar hung on the potential changing geopolitical boundaries of Europe. Things seemed clear next to the Rhine River: The Rheinberg soda plant and the Borth salt mining plant fell in the British zone. The Wyhlen plant, close to Switzerland, belonged to the French zone. But the overwhelming majority of DSW industrial capacities were located further east, in Saxony and Anhalt. Bernburg and Eisenach were liberated by the American troops on 16 April 1945. What the workers and the populations did not know, however, was that the occupation of the region would be taken over by the Russians in early July 1945 according to the provisions of the Yalta Conference. An indescribable confusion thus reigned in and around the Bernburg factory between the liberation and the transfer to the Soviet occupying authorities. On 22 June 1945, the American Counter Intelligence Corps initiated the "exfiltration" to western zones of occupation of most of DSW's higher managers and their families. For Ernest-John Solvay and René Boël alike, the conditions in which this operation took place were rather mysterious; it amounted to a form of kidnapping. Obviously, Brussels had not ordered this awkward rescue mission. It suddenly drew to the American zone forty DSW personnel members who were not authorized to be appointed to DSW Western plants.

---

[6] This and following sections draw on Bertrams, Coupain, and Homburg, *Solvay*, chap. 13.

Figure 7.2. The Potsdam Conference in Germany (July 1945) during which representatives of the "Big Three" powers gathered to decide how to administer punishment to the defeated Nazi Germany. U.S. President Harry S. Truman is in the left foreground. British prime minister Winston Churchill is seated at the upper left of table. Clement Attlee is seated two to the right of him. Soviet leader Josef Stalin is at the upper right of table. Soviet foreign minister Vyacheslav Molotov is to his immediate left. (Harry S. Truman Library)

They were confined in Dornheim, a village south of Frankfurt, and had to wait there for their "evaluation" to take place – that is, their certificate of denazification provided by the American military authorities. This could drag on for months, sometimes for years.

The transfer of DSW higher management to the West contributed to the disruption of the chain of command, which had already been seriously affected. René Boël took special care of this issue. Helmut Eilsberger remained in charge, but Carl Adolf Clemm, who had been the wartime *Verwalter* (administrator) of DSW, was not recognized by the Allied authorities. Shortly before his transfer to Dornheim, he had entrusted Otto Bökelmann as head of Bernburg. Close to democratic-liberal political groups – his brother-in-law had been killed by the Nazis – Bökelmann was well known within antifascist circles. On the other hand, his relations with the Russians proved complex. When Bernburg plant was officially sequestrated on 17 April 1946, just a year after its liberation, nobody really knew how things would evolve in the Soviet zone. Worries concentrated above all on the manner in which reparations were to be extracted and on the level of administrative autonomy. The Potsdam Agreement had already given the Russians a generous leeway. Some observers even spoke of a virtual carte blanche in terms of reparations. By late spring 1946, the four-power authorities decided that plants in excess of the capacities stipulated were "supposedly earmarked for dismantling and distribution to Germany's reparation creditors."[7] By and large, dismantling did not take place, or did to a very limited extent, in the American, British, and French occupation zones. What did take place, however, was a procedure of technology transfer, mostly unwilling, from German to Allied industry. In this context, Solvay benefited from hydrogen peroxide processes, which it constantly developed and improved throughout the 1950s.[8] The politics of dismantling applied in the east to a large extent. Soviet forces had slowly but surely started to dismantle Bernburg facilities as early as November 1945. With the new regulations, the pace of dismantling quickened. At Bernburg plant, it was ultimately completed on 23 April 1948. Nine days earlier, the General Assembly meeting of DSW officially approved the transfer of DSW's headquarters to Solingen, located 70 kilometers south of Rheinberg and at a close distance to Düsseldorf, Leverkusen, and other key locations of the German chemical industry (Bayer and Henkel, for instance). Everyone understood what it meant: the separation of

---

[7] Charles S. Maier, "'Issue Then Is Germany and with It Future of Europe'," in C. Maier and G. Bischof (eds.), *The Marshall Plan and Germany*, New York-Oxford, Berg, 1991, 19.

[8] Bertrams, Coupain, and Homburg, *Solvay*, chap. 15.

Germany was no longer a threat. The creation of an independent Eastern Germany under Soviet influence became real and tangible, although the German Democratic Republic (GDR) was officially established only on 7 October 1949. Meanwhile, a similar fate was about to sweep the rest of Central and Eastern Europe.

THE DIVISION OF EUROPE AND ITS HARSH CONSEQUENCES

In many ways, the years 1945–7 were similar to 1919–20. Political instability, social upheavals, and multilateral diplomatic engineering dominated the European scene. But this should not give way to hasty conclusions; there was indeed a huge difference between the two periods. Whereas Europe's first postwar ended with the (violent) taming of communism, the second postwar gave way to two modes of enduring subordination: in the East to the Soviet Union, in the West to the United States.[9] This double hegemony over Europe was echoed by a stark antagonism between the two former Allies. It found a physical metaphor in the "Iron Curtain," which eventually crossed Europe and divided it into two separate realities for more than forty years. The Iron Curtain expression was coined by Winston Churchill in a prescient speech he gave at Westminster College in Fulton, Missouri, on 5 March 1946.[10] At the time of Churchill's speech, the building of the Soviet Empire was scarcely complete. It was an ongoing process among others. What became increasingly evident, however, was that countries from Central and Eastern Europe were eager to implement measures of nationalization of their strategic industries. Quite true, they were not the only ones to do so. Laws of nationalization were adopted in France and Italy as a means to promote a State-interventionist economic policy compatible with liberalism. But in Central and Eastern Europe, the degree of nationalization reached unprecedented levels. It was the instrument of an overarching planned economy rather than the

[9] Mazower, *Dark Continent*, 216–17.
[10] "From Stettin in the Baltic to Trieste in the Adriatic an "iron curtain" has descended across the continent. Behind that line lie all the capitals of the ancient States of Central and Eastern Europe.... [They] lie in what I must call the Soviet sphere, and all are subject, in one form or another, not only to Soviet influence but to a very high and in some cases increasing measure of control from Moscow."

expression of the "mixed economy" in vogue in Western Europe. Czechoslovakia set the stage with the introduction of a decree-law of nationalization on 24 October 1945. It was soon followed by Poland, Yugoslavia, Albania, and Bulgaria in 1946. Rumania, Hungary, and, much later on, the GDR, rounded out the process.[11]

As we know, the Solvay group was firmly rooted east of the Iron Curtain. With the exception of Bulgaria and Albania, it controlled chemical companies in all the countries that became "Soviet satellites" after 1948. Faced with the sudden expropriation of their plants, the *gérants* reacted with intransigence. "What counts for us," noted René Boël in January 1947, "is the ownership of our plants [in Central and Eastern Europe] and we should strive at keeping this line as much as we can, especially in the field of soda ash."[12] Yet keeping this line proved difficult as states were assuming ownership of most of the major branches of industry. Widespread nationalization implied that private businesses like Solvay were declared illegal and that severe limits were imposed on the right of individuals to produce for the market.[13] Getting back its eastern factories was thus hardly an option for the Solvay group. The only viable alternative that remained was compensation. Everywhere in the West, joint industrial committees engaged in painstaking negotiations with the authorities from the so-called People's Republics to claim the financial counterpart of their possessions. The enduring efforts scarcely paid off. There were two exceptions. Solvay first obtained a fixed-rated compensation from Yugoslavia, which accounted for 85 percent of its majority holding in the Lukavac soda plant and electrolytic unit, as early as October 1948. It was much later, however, that the Czech deal was solved. Only in 1984 did the Verein Aussig agree with Solvay to complete the terms of the negotiation. Elsewhere, the compensation scheme was irritating, frustrating, and almost fruitless. Overall, it amounted to the loss of fifteen plants in Central and Eastern Europe.

---

[11] Berend, *Economic History*, 154–56.
[12] ACS, 1001–37-1, "Réunion du 3 janvier 1947" with R. Boël, H. Delwart, R. Kirkpatrick, G. Janson, M. Ebrant, and Éd. Swolfs, 10 January 1947.
[13] Barry Eichengreen, *The European Economy since 1945. Coordinated Capitalism and Beyond*, Princeton, Princeton University Press, 2007, 134–35.

Figure 7.3. The "Solvay Trial" at Bernburg, 14 December 1950. This show trial was filed by the GDR authorities against local Solvay managers for economic crimes. It was mainly a pretext for the seizure of the East German Solvay plants by the communist regime (Photographer: Walter Heilig. Bundesarchiv, Berlin)

Among them, the industrial capacities located in Eastern Germany were by far the most important for Solvay. According to prewar and wartime figures, the factories that were in GDR accounted for roughly 75 percent of DSW's workforce, 55.1 percent of the overall production of soda ash, and 81 percent of caustic soda. There was no electrolysis in Western Germany, nor any potassium salts works or potassium and lignite mines. For more than fifty years, Bernburg had been much more than DSW's administrative headquarters. It had been the heartland of Solvay's successful development in Central and Eastern Europe, as well as the Solvay Group's biggest factory. After its dismantling under the Soviet occupation, a portion of Bernburg's soda ash facilities was rebuilt by 1952. On 5 May 1953 as a means to signify a radically new departure, the Eastern German authorities renamed the factory as the Sodawerke Karl Marx. This was not strictly speaking a measure of nationalization, although it came close to it. The GDR authorities instead talked about "supervision." The subtle distinction proved crucial when, in 1991, Solvay could eventually take Bernburg back, without payment (see Chapter 9).[14] Yet all this seemed rather anecdotal compared with an earlier, and infamous, event. In December 1950 and June 1951, the GDR regime had organized two large-scale "Solvay trials," quite analogous to Moscow's show-case trials, which led to the condemnation as "economic criminals" of several Bernburg plant managers. A fiction film produced by the State titled *Geheimakten Solvay* (1953) drew on these trials as a way to deliver a quite dull, propaganda-style sabotage story. What was quite real, however, was the fact that nine innocents had been sentenced to penalties varying between two and ten years' imprisonment. They could only rebut the central piece of their accusation – namely, that they had disclosed information about the exchange of capital between DSW and I.G. Farben. The defendants had no clue about this financial agreement, which was concluded in 1924. Then again, the judge deliberately did not take into account the evidence sent by Brussels and Solingen, which proved their innocence.

WESTERN EUROPE: RECOVERY, RECONSTRUCTION, INTEGRATION

With the loss of its eastern European factories, the Solvay group was severely wounded.[15] Nevertheless, the *gérants* did not lament passively on the transformation of the international geopolitical situation. Some of them still remembered the loss of the Russian plants after the First World War and the subsequent development of large chemical companies in the 1920s. More important, all of them had experienced, albeit in different ways and places, the dramatic events of the war. A common conviction resulted from this personal background: If it wanted to reach its centenary, Solvay & Cie had to change in a rapidly changing world. The second postwar

---

[14] Bertrams, Coupain, and Homburg, *Solvay*, chap. 20.
[15] Interestingly, the premises were not officially nationalized until 1972. Only after that year were contacts established with the authorities of the GDR in view of compensation, which never came.

ushered in a period of profound social and political transformations, which determined the two or three following decades. In sharp contrast with the prewar era, liberal democracy now seemed the indisputable and consensual political horizon of all Western European countries that sought stability.[16] Also crucial was the enactment of a series of long-awaited social reforms, which went beyond the scope of curtailed measures of labor welfare. These legislations now formed the foundations of a full-fledge system of "social security." The safety net was organized by the State, not by private companies; it was designed to support all citizens "from the cradle to the grave" irrespective of their working conditions. Evidently, this sudden transition from warfare to welfare was intended to foster the reconciliation of people with their State and to provide them a feeling of normality.[17]

The expanding role of the State was also visible in the institutional engineering of Western Europe's political economy. Drawing on prewar and wartime experiments, planners became fairly successful in crafting an original economy mix between Eastern Europe's State dirigisme and U.S.-style market economy. Special ministries, agencies, committees, and other intergovernmental offices now filled the administrative landscape as a means to promote a coordinated version of capitalism. In France, Jean Monnet, a former legal adviser of Solvay's interests in the United States who later became a skillful diplomat for Charles de Gaulle, became head of the General Planning Commissariat in January 1946.[18] Its overly ambitious task was to set the French economy on its way to reconstruction and modernization by emphasizing capacity in basic and strategic industries. In this sense, the Monnet Plan complemented another endeavor of planning that dealt with six European countries. The European Coal and Steel Community, which was officially presented by French foreign minister Robert Schuman on 9 May 1950 (now still commemorated

---

[16] Martin Conway, "The Rise and Fall of Western Europe's Democratic Age, 1945–73," *Contemporary European History*, 2004 (13), 67–88.
[17] See Richard Bessel and Dirk Schumann (eds.), *Life after Death. Approaches to a Cultural and Social History of Europe During the 1940s and 1950s*, Cambridge, Cambridge University Press, 2003.
[18] Philippe Mioche, "Jean Monnet, homme d'affaires à la lumière de nouvelles archives," *Parlement(s)*, 2007/3, 55–72.

as Europe Day by the European Union's members), largely rested on Monnet's confidential draft for Franco-German reconciliation. The pooling project of coal and steel production governed by a daring supranational High Authority was seen by Monnet as a first step toward an overarching political integration of European societies – a design that ultimately did not materialize after the failure of the European Defense Community in 1954.[19] However, Monnet's undertaking had found a staunch ally in René Boël, whose wartime experience had persuaded him of the need to bring European countries on the path to stronger economic cooperation. From 1951 until 1981, Boël was the president of the European League for Economic Cooperation, a think tank that strove for standardization among Western European markets and legal systems.[20]

THE AMERICAN LEADERSHIP BETWEEN CONSTRAINTS AND SEDUCTION

In Boël's view, it was obvious that the United States would orchestrate Western Europe's integration. This resulted from his personal experience during the war, but there was more at stake in this respect. Contrary to the first postwar, the aftermath of the Second World War had largely confirmed the United States' ambition to become a major political actor in the world at large, and in Europe and Japan in particular. Following the immediate postwar military-style occupation in Germany, the United States did not withdraw from Europe. Quite the contrary, it proposed the European Recovery Program in June 1947, better known as the Marshall Plan. Using a strictly economic approach, the Marshall Plan provided some $13 billion in grants and loans, which allowed its European beneficiaries to buy American materials and equipment. This kind of economic assistance between Europe and the United States was nothing new. Yet unlike previous relief-based initiatives, the Marshall Plan addressed many overlapping obstacles with one blow. It solved

---

[19] Eichengreen, *European Economy*, 167–9.
[20] See Michel Dumoulin and Anne-Myriam Dutrieue, *La Ligue Européenne de Coopération Économique (1946–1981). Un groupe d'étude et de pression dans la construction européenne*, Bern, Peter Lang, 1993.

Figure 7.4. Marshall Plan Funds to West Germany totaled $1,390,600 and intended to rise the country from the ashes of defeat, as illustrated by this propaganda photograph showing a worker in West Berlin. Already a year before the end of the program in 1951, West Germany had surpassed its prewar industrial production level. (U.S. National Archives and Records Administration)

trade deficits, countered inflation, and restored price liberalization, but above all the impact of the Marshall Plan was political, if not ideological, in the context of the rising Cold War. Beneficiaries had to choose their camp: a government-planned or market-oriented economy.[21]

[21] Alan S. Milward, *The Reconstruction of Western Europe, 1945–1951*, London, Methuen, 1984, 56–61.

Thus, Western Europe made its entrance into the age of affluence through the American doorway. The United States, needless to say, epitomized the virtuous circle of the so-called Fordist model of economy that West European countries were eager to adapt – rather than adopt – to local circumstances. Its recipe consisted of uninterrupted growth, mass production, increasing productivity, higher wages, full employment, and mass consumption.[22] By the same token, many European industrialists were staring at the ever-expanding prospects of the U.S. market. Investing in the United States, however, was no bed of roses for foreign companies. Looking back at more than sixty years of uneasy relations with its American partners, Solvay & Cie knew this too well. In the chemical industry in particular, the involvement of American military authorities in breaking up of the German conglomerate I.G. Farben had been more far-reaching than the other Allied occupying powers.[23] The disentanglement of the German trust gave way to the (re)birth of mighty European competitors (BASF, Bayer, and Hoechst). The United States was equally mighty in the field of antitrust legislations. In March 1944, a U.S. district court in New York had ruled that the U.S. Alkali Export Association (Alkasso) could no longer operate in the United States under the Sherman Antitrust Act of 1890. Alkasso had been set up in 1919 as a means to make selling arrangements among the largest ammonia-soda manufacturers present on the North American continent, which included the Solvay Process Company and ICI.[24] The impact of the United States on foreign businesses was thus intense.

Yet the lure of the market prevailed over other constraints; the United States was the place to be after the Second World War. Impressed and inspired by what he saw during his stay in New

---

[22] For a balanced view, see Charles F. Sabel and Jonathan Zeitlin (eds.), *World of Possibilities: Flexibility and Mass Production in Western Industrialization*, New York/Cambridge, Cambridge University Press, 1997.
[23] Raymond G. Stokes, *Divide and Prosper. The Heirs of I.G. Farben under Allied Authority*, Berkeley, University of California Press, 1988.
[24] Glasscock, *Commercial History*, 45–52. Alkasso is thus not the predecessor of Ansac, which was created in 1984 as a platform for international sales and distribution of natural soda ash. Solvay became de facto member of Ansac after the acquisition of the Green River plant of Tenneco in 1992.

York as "*gérant*-in-exile," René Boël wished to seize the opportunity. The expected loss of Eastern European factories prompted him to look to the West. It must be said that this "Americanization" concerned the Americas, broadly conceived. In other words, it also included Brazil, where René Boël had sown promising seeds, which still needed to grow and blossom.[25] The United States remained a special target, however. Underlying Boël's intention was the desire to diversify Solvay's financial investment in the United States, the bulk of which concentrated on shares of Allied Chemical and Dye Corp., as well as to have a firm bridgehead in the booming American organic chemical industry.[26] His radar was alerted by the family business Wyandotte Chemicals Corporation, previously known as Michigan Alkali, which had some history with Solvay & Cie. For several reasons, Boël intended to move forward and opted for a (partial) takeover of Wyandotte. Preliminary negotiations had been quite successful, but when the time was ripe to complete the deal in 1953, it had collapsed. The reasons for the failure hinged as much on the main American shareholders' opposition to concluding the agreement as on Ernest-John Solvay's reluctance to carry out the program.[27] Apparently, Boël's colleague at the *gérance* had not overcome his skepticism toward American business milieus, which stemmed from Solvay & Cie's tumultuous relations with Allied during the Weber era. As it happened, it would be Ernest-John's son, Jacques Solvay (1920–2010), who would ultimately succeed in taking over a major American chemical activity, with the acquisition of the polyolefins business of Celanese in 1974, which included production facilities in Deer Park, Texas (see Chapter 8). In the meantime, however, Solvay & Cie had entered the realm of diversification through another gateway.

---

[25] Interestingly, the counterpart of Solvay & Cie's I.G. Farben shares was partly redirected to cover expansion into Brazil. See ACS, "Politique industrielle du Groupe Solvay," Interview of Comte Boël (16 January 1987), 19 January 1987.
[26] Unless otherwise noted, the section draws on Bertrams, Coupain, and Homburg, *Solvay*, chap. 18.
[27] Wyandotte's main shareholders were the Ford family, which became active in the production of plate glass in Ohio both before and after setting up the Libbey-Owens-Ford Glass Company in 1930. These Fords are not related to the well-known automobile family business.

Figure 7.5. Vinyl records made of PVC symbolize entrance into the culture of mass consumption and the plastics era. Solvay took part in this trend by diversifying its activities toward the production of plastics. (Mary Evans Picture Library / Classic Stock / Coleman)

EMBARKING ON THE PLASTIC DRIVE

Changes in international business and politics altered the political economy of energy production and consumption. Whereas Europe used relatively little petroleum in 1938, by 1955 it accounted for 20 percent of the continent's total energy consumption. The figure rose to 45 percent in 1964.[28] The chemical industry was both a witness and a driver of this shift. The impact was particularly impressive when polymers such as polyethylene evolved into major industries. Its huge production entailed a radical change of chemical feedstocks. As an expert put it, "regardless of the fact that Europe's chemical industry was for a long time more advanced than that in the United States, the future of organic chemicals was going to be related to petroleum, not to coal."[29] Quite true, the real breakthrough from carbochemicals to petrochemicals took place in the United States where synthetic fibers and polymer products had been brought to market in the 1930s. This enabled companies such as Du Pont and Union Carbide to enter the Second World War with an innovator's advantage and reap the harvest of the petrochemical postwar boom.[30] These products were the outcomes of a long research and industrial process, which stretched from fundamental research to development-oriented chemical engineering. Nylon, for instance, a highly symbolic product of postwar material culture invented at Du Pont's Research Laboratory in 1935, has an extensive history that went back to the Bakelite polymer resin obtained by the Belgian American chemist Leo Baekeland in 1907.[31] The same was true for polyvinyl chloride, usually referred to as PVC.[32] It was first

---

[28] Raymond G. Stokes, *Opting for Oil. The Political Economy of Technological Change in the West German Chemical Industry, 1945–1961*, Cambridge/New York, Cambridge University Press, 1994, 95–6.

[29] Peter H. Spitz, *Petrochemicals: The Rise of an Industry*, New York, John Wiley & Sons, 1988, xiii.

[30] Alfred D. Chandler, *Shaping the Industrial Century. The Remarkable Story of the Evolution of the Modern Chemical and Pharmaceutical Industries*, Cambridge, MA/London, Harvard University Press, 2005, 46–7, 72–3.

[31] Hounshell and Smith, *Science and Corporate Strategy*, 257–74; Ndiaye, *Nylon and Bombs*, 92–5.

[32] This and the following sections draw largely on Bertrams, Coupain, and Homburg, *Solvay*, chap. 15.

discovered in 1912, industrially improved during the interwar years, and gained a momentum after the Second World War. Only by then did it truly meet the cultural requirements of the postwar consumer society. From packaging to automobile cables and "vinyl" music records, PVC was used for a wide variety of applications. It was the versatile all-around plastic par excellence. It especially garnered attention in domestic life because it was used for commodities such as children's raincoats, garden hoses, women's purses, and, last but not least, Mattel's flagship Barbie dolls, first introduced in 1959.[33]

Producing plastic dolls was perhaps not Solvay's initial intention when the war drew to a close, but the company ended up being a leader in PVC production in continental Europe by the end of the 1960s. This was both a spectacular and an unexpected achievement. Most important, however, it marked Solvay & Cie's striking endorsement of the diversification strategy. The issue had been debated time and again within the *gérance* but had been always discarded or downplayed by a majority of *gérants*. Although various research and development (R&D) activities in organic chemistry, focusing on chlorine-based products in particular, had been realized since the mid-1930s, World War II represented a watershed in terms of industrial strategy. Solvay & Cie would no longer be a strictly and exclusively alkali-centered company after 1945. This did not mean that the production of soda ash and caustic soda was abandoned nor that the company's endeavors had to be started from scratch. Quite the contrary, the move toward chlorine products, like plastics, grew with the explicit objective of defending and protecting the caustic soda market. Hence, new R&D programs were carried out at the company's Central Laboratory in Ixelles, as well as in the laboratories of the Jemeppe, Tavaux, and Zurzach plants. In 1952, the breadth of Solvay & Cie's postwar research agenda prompted the transfer of R&D activities from the old Ixelles premises to a new complex built at Neder-over-Heembeek (NOH), in the north of Brussels. The basic idea was to orient and enhance research on existing production lines, especially with the by-products of electrolysis. Even before the war, largely under René Boël's impulsion, the

---

[33] Jeffrey L. Meikle, *American Plastic. A Cultural History*, New Brunswick, Rutgers University Press, 1995, 186–7.

company made important investments in electrolytic chemistry.[34] At that time, however, the main purpose was the alternative production of caustic soda. Whatever their ultimate goal, these interwar experiments would later prove to be instrumental. As noted by Ernst Homburg, "without that background it would have been almost impossible to become a successful producer of PVC."[35]

DIVERSIFYING THE DIVERSIFICATION

A salient expression of Solvay & Cie's commitment to product diversification was that it also opted to diversify its strategy. Parallel to its in-house R&D activities, Solvay decided to rely on its long-time industrial partner and international ally, the British multinational ICI. It was a clever move for at least four reasons. First, ICI held major legal international contracts that were complementary to those of Solvay in terms of world market sharing. Second, the war may not have transformed ICI into the biggest and most profitable of British companies, but the overall performance of British big business in the 1950s (and in the following forty years) made ICI a central actor in the world chemical industry.[36] Third, ICI was extremely engaged in research. In 1945, it became the largest British contributor in terms of R&D expenditures and turned out to be the biggest private employer of qualified research personnel – no less than 1,396 engineers, chemists, and technicians worked in its laboratories.[37] Last but not least, ICI knew a lot about PVC. By 1945, while the large-scale industrialization of PVC was still much limited in Europe, it was a proven technology for the British company. Thus, by teaming up with ICI, Solvay chose a fine partner and enjoyed strong momentum. After the initial stages of technology exchange, which began in November 1945 (here again, in line with a long-established practice of transfer of knowledge and know-how), both companies decided to launch a more ambitious

---

[34] See Bertrams, Coupain, and Homburg, *Solvay*, chap. 11.
[35] Bertrams, Coupain, and Homburg, *Solvay*, 338.
[36] Cassis, *Big Business*, 95–9.
[37] David Edgerton and Sally Horrocks, "British Industrial Research and Development before 1945," *The Economic History Review*, 47 (2), 1994, 223–6.

project. Instead of ensuring the transfer of nonexclusive licenses, an agreement was concluded on 15 October 1949 that set up a joint-venture business partnership for the production of PVC. Solvic, as it was called, was a remarkable success. Thanks to the partnership, factories were in motion even before the official conclusion of the agreement. With subsidiaries in almost every European country and even Brazil, Solvic was the driving force of Solvay's international expansion in PVC. Resting on this joint-partnership performance, Solvay & Cie's in-house research departments considerably improved PVC products. The most famous achievement was the PVC plastic bottle, first manufactured by a research team at Tavaux plant in 1963. Under the impulse of Solvay's *gérant* Paul Washer (b. in 1922), it was later designed for use by French mineral water producers, advancing the shift from glass to plastic for foodstuff applications.[38]

Another polymer, polyethylene, aroused international attention by the late 1940s. Lighter, more flexible, and less permanent than PVC, polyethylene had been invented by a team of ICI scientists in the mid-1930s. After the war, it became the symbol of a plentiful society whose culture was "shapeless, ever-changing, impermanent, even ephemeral."[39] Not surprisingly, all kinds of chemical businesses sought to get a piece of this pie, and Solvay was no exception. Yet until ICI's British patent expired in 1956, companies that wished to take part in the "polyethylene bonanza" either had to purchase the license from ICI or develop alternative technologies.[40] As it happened, Solvay & Cie first opted for the former solution but was forced to move to the latter as obstacles unexpectedly appeared. The *gérants* stuck firmly to their conviction that polyethylene brought decisive competitive advantages. There was an essential caveat, however. Epitomizing the shift from coal to oil, polyethylene demanded an exclusive supply of ethylene – a raw material that Solvay could not provide nor control. This represented not only a radical change in the company's practices; it was a

[38] Bertrams, Coupain, and Homburg, *Solvay*, chap. 16.
[39] Meikle, *American Plastic*, 177.
[40] Stokes, *Opting for Oil*, 121.

deviation from the founding father's legacy to have the upper hand and the ultimate say on feedstocks. The *gérants* were fully aware of this but decided it was a risk worth taking. Several methods were explored to overcome this disadvantage, but none ultimately proved convincing. In Brazil, for instance, the polyethylene unit built at the Elclor plant was supplied on ethylene made of homegrown sugar cane. At the Rosignano plant in Italy, a cracker, which enabled the breakdown of large alkalenes into smaller and more useful molecules, was purchased from the Italian firm Montecatini. It turned out to be a failure.[41] Then again, these drawbacks were partly compensated for by Solvay's push for in-house innovation. In particular, a brand-new catalyst for the production of polypropylene was discovered in the research premises of Neder-over-Heembeek in 1969. It yielded significant income from licensing.

At the turn of the 1960s, a "new" Solvay & Cie was on track. To a large extent, however, it was the "old" Solvay & Cie's real heir. The challenge of the plastic revolution had been taken up alongside the expansion of the company's "historic" alkali activities. Original business ventures such as Solvic stemmed from long-standing industrial partnerships and agreements. Yet does it all come down to the "old-wine-in-new-bottles" perspective? It certainly does not (and not only because such bottles were not made of PVC). Take the impetus for R&D, for instance: It was both strong and new. With the exception of some organic chemicals, Solvay began research in catalysis, polymers, and materials after the Second World War. The domain of hydrogen peroxide, already mentioned, is another case in point. Thanks to a series of developmental improvements realized in its various research sites throughout the 1950s, Solvay attained leadership in this area that had been unknown to the company before the war. Its position was further improved in this profitable domain after concluding a partnership with the British company Laporte. It led to the creation in 1971 of the joint-venture Interox, with a worldwide dissemination of plants.

By 1947–8, few would have bet on Solvay & Cie's survival fifteen years later, let alone on its ability to still rank among the

---

[41] Bertrams, Coupain, and Homburg, *Solvay*, chap. 16.

world's largest chemical companies. Commemorating its centennial anniversary in December 1963, the family business had many reasons to celebrate. Diversification was finally recognized as a viable strategy. The company was forward-looking, freed from the burdens of the past. The international economy was in full swing. The sky was the limit to growth and expansion. But how long could such good fortune last?

Figure 8.1. Drug manufacturing at Kali-Chemie (early 1990s). This German subsidiary was, together with Duphar and Salsbury, one of the pillars on which Solvay based its move into life sciences. (Solvay Archives)

8

Recession and the Biochemical Impulse

> The world economic situation should not be seen with overstated pessimism. It should be seen with lucidity and realism. It concerns a serious accident, not a catastrophe. We should strive to contain any form of dramatization. Bringing up the specter of the crisis of 1929–1930 is not sound.
>
> Max Nokin (March 1975)[1]

> We are now, though we only dimly begin to realize the fact, in the opening stages of the Biological Revolution – a twentieth-century revolution which will affect human life far more profoundly than the great Mechanical Revolution of the nineteenth-century or the Technological Revolution through which we are now passing.
>
> Gordon Rattray Taylor (1968)[2]

THE END OF THE "GOLDEN AGE"

In 1979, the French economist Jean Fourastié wrote a book that became influential above all for its well-chosen title: *The Glorious Thirty, or the Invisible Revolution from 1946–1975*. In hindsight, it was indeed obvious that the thirty years following the Second World

---

[1] Quoted in Jacques Moden and Jean Sloover, *Le patronat belge. Discours et idéologie, 1973–1980*, Brussels, CRISP, 1980, 152. Max Nokin (1907–96) was head of the industrial holding Société Générale de Belgique from 1961 to 1974. See Ginette Kurgan-van Hentenryk et al. (eds.), *Dictionnaire des patrons*, 485–7.

[2] G. Rattray Taylor, *The Biological Time Bomb*, New York, New American Library, 1968, 13 quoted in Robert Bud, *The Uses of Life. A History of Biotechnology*, Cambridge, Cambridge University Press, 1993, 163.

War were a period of growth, prosperity, and social improvement. But from the perspective of 1979 (or later), it seemed even more obvious that this "Golden Age" was definitely over, a ghost of the past. The first signs of the economic slowdown that affected the world's economy were already visible at the end of the 1960s, with growing anxiety concerning the international monetary system. Yet it was not until 1975 that the perception of a structural crisis gained acceptance in various milieus. Continuous inflation, reduction of growth, eroded competitiveness, and decreasing productivity were tightly linked to high levels of unemployment and social discontent. The conditions of the postwar labor market, especially the connection between better wages and social harmony, no longer existed. As plants and facilities were shut down, strikes and demonstrations broke out regularly. In the United States and Western Europe, "most of the radicals of the Sixties, like their followers, abandoned 'the Revolution' and worried instead about their job prospects."[3] This proved inevitable as unemployment rose to levels that approached those of the dreadful 1930s. With falling employment security, workers requested and obtained some forms of unemployment compensation. This put extra pressure on public finances, which were already in dubious shape.

By the late 1970s, conservative governments, such as that led by the United Kingdom's prime minister Margaret Thatcher, introduced a series of measures that tightened the State's welfare legacy, transformed the labor market, and deregulated the financial environment but could hardly solve the endemic recession.[4] In the United States, President Ronald Reagan adopted a fiscal policy that echoed the Thatcherite experiment. It was based on lower taxes that stimulated so-called supply-side economics as opposed to demand-based growth, which had been set into motion since the mid-1940s. Margaret Thatcher and Ronald Reagan's economic policies rested on the same neo-liberal premises, which tended to alleviate the importance of the State in the economy and favor growth by risk-taking innovation. Considering the institutional and symbolic weight of the welfare state, this was akin to a public management revolution. Elsewhere in Western Europe, most notably in François

---

[3] Judt, *Postwar*, 453.
[4] Eichengreen, *European Economy*, 272–82.

Figure 8.2. After the oil crisis of 1973 the price per barrel multiplied by four and shortages were looming. This event brought an end to low-cost energy. (Mary Evans Picture Library / Classic Stock / H. Armstrong Roberts)

Mitterrand's France, there were attempts to counter the neo-liberal wave and extend the postwar social blueprint, but enduring realization fell short. The ongoing economic stress forced national governments to curb their ambitious social promises and withdraw into austerity and *rigueur*.[5] The early 1980s thus represented a watershed in terms of economic policy. Neo-liberal values not only held sway in the economic agenda, they had an impact on the social

---

[5] Mark Mazower, *Dark Continent*, 342–3.

vision at large. As prime minister Thatcher once famously asked: "Who is society? There is no such thing! There are individual men and women and there are families."[6]

Against this gloomy backdrop, companies were hoping for the best but preparing for the worst. The outbreak of two oil crises in ten years' time, in 1973–5 and in 1979–81, ushered in a period of uncertainty. Blind optimism had given way to overproduction. The recession hit hard key sectors of the industry, all of which had been thriving during the booming years of the mass-consumption society a decade earlier. The chemical industry was no exception. Marked by an increasing reliance on petrochemical feedstock combined with the trend of higher wages inherited from the postwar social contract, chemical companies were facing rising costs, especially in Europe and in the United States. Furthermore, the overcapacity in production of basic chemical commodities that resulted from market saturation led to a continuous drop in sales prices. How could this be solved? Anxiety and fear were looming as industrialists compared the slump to the Great Depression (see the opening quote of this chapter). However, as years of recession went by, industrialized countries exhibited signs of unexpected resilience. After a series of efforts that led to structural adjustments, it appeared that the kind of systemic collapse triggered by the market stock crash of 1929 had been averted. Industrial capitalism had thus survived its second major shock in half a century. Then again, such a statement is easy to make in retrospect. Historians' expertise only throws light on past events. Hence, we need to take a closer look at these years as though they were unfolding in real time.

FIRST STRATEGY, THEN STRUCTURE: SOLVAY & CIE BECOMES A PUBLIC COMPANY (1967)

By the mid-1960s, under the impulse of *gérants* such as René Boël and Henri Delwart (1902–87) in the first stage, then Jacques Solvay and Paul Washer in the second stage, Solvay & Cie had successfully

[6] Quoted in Bernard Wasserstein, *Barbarism and Civilization. A History of Europe in Our Time*, Oxford/New York, Oxford University Press, 2007, 636.

taken up the long-awaited challenge of product diversification.[7] Its world leadership in the historical branch of alkalis was uncontested, especially in continental Europe where its market shares were often above 60 percent. Its position in industrial activities deriving from electrolytic production, which had resulted in the production of PVC and peroxides, recorded continuous growth. Finally, the decision to foster backward and forward integration, from the production of polyethylene to plastics processing, was logical as a means to capture an entire technological system. Solvay & Cie, as a result, was a rapidly changing company as far as industrial strategy was concerned. Its structure, however, still largely corresponded to the organizational outlook designed by the founding fathers in the nineteenth century. For obvious reasons, the attachment of the family business to the principle of financial and corporate independence was both profound and sincere. At the beginning of Solvay's industrial adventure, it was believed that the legal form that suited best this autonomy was that of a partnership company. Such a legal configuration had the advantage of being clear and simple: The firm was theoretically organized between silent partners and managing partners; the former brought the capital, the latter ran the business. Within this framework, the practical arrangements, just like the company's day-to-day management, witnessed many changes through the years. But one thing was clear from the outset: The legal form of partnership was not considered permanent. Quite the contrary, Solvay & Cie's bylaws had rendered a change of legal regime relatively easy to accomplish.

The transformation of Solvay & Cie into a joint-stock company was, therefore, an ongoing issue. One could say the debate was as old as the company itself. Motivations to promote the adoption of a new legal disposition differed in their context, though. In the mid-1880s, for instance, the issue was intensely discussed among associates and *gérants* to facilitate capital circulation among shareholders.[8] In 1928, by contrast, when Armand Solvay

---

[7] This section draws largely on Bertrams, Coupain, and Homburg, *Solvay*, chap. 17.
[8] See Bertrams, Coupain, and Homburg, *Solvay*, chap. 3.

supported the transformation into a joint-stock company, his main concern was to avoid a troubled succession to the head of the business after his death. Impressed by the organizational flexibility displayed by Allied Chemical and ICI, Armand proposed relying more systematically on outside professionals. "For the leading positions," he told his colleagues, "a company like ours needs to choose 'aces'; if we can recruit these aces among ourselves, that's just fine; but if we can't, we need to find them where they are."[9] In the 1960s, managerial issues were not at the top of the agenda. The priority was to reform the financial architecture and, more precisely, address Solvay & Cie's need to fulfill its ambitious investment program through improved access to capital markets.

It would be unfair to say that inertia had prevailed before this point. Important measures had been taken to inject extra cash into the company in the meantime. The choice to sell significant parts of Solvay's holdings in ICI and the decision to cut short the recruitment of new personnel and even to close down several soda ash factories (Bayonne and Monfalcone) or caustification units (Zurzach, Dombasle, and Rosignano) were made during the 1960s. But this was a change in degree, whereas a change in nature was the objective. It is in this context that the transformation into a joint-stock company came to the fore once again. As a result of Ernest-John Solvay's opposition, the issue had stalled for a while. But when René Boël succeeded him at the chairmanship of the *gérance* on 1 January 1964, the process went into full swing. After technical obstacles were successfully overcome, a decisive General Assembly meeting took place on 12 June 1967. The green light was given for the transformation of the partnership into a joint-stock company. After 104 years of existence, Solvay & Cie ultimately became a public company. The so-called initial public offering took place in November 1967. Considering the secretive dimension of the company, among other salient characteristics, this public shift was a genuinely historic turn, but it had been a carefully prepared move as well. Not surprisingly, the new legal status was creating as many challenges as it had solved constraints. There were many obstacles

---

[9] ACS, "Aperçu historique sur la Gérance de Solvay & Cie et considérations relatives à une transformation éventuelle de notre Société en Société anonyme," by Armand Solvay, 9 January 1928.

along the way. Capital absorption was one of them. To avoid a reckless dispersion of shares that would have entailed unfriendly takeover bids, René Boël, Paul Washer, and Solvay's legal counsel Henri Lévy-Morelle (1920–2012) strove to reach an agreement with financial and legal experts. After all, historical shareholders had accepted the legal transformation under the condition that the company remain a family company. Corporate control was by no means negotiable. An original solution was eventually found out: There would not be one Solvay share, but three types of shares (A, B, and C), with the two last categories being registered shares and therefore subject to specific restrictions. The majority of Solvay shares, therefore, could remain in the hands of the shareholding families. A second step was reached when the threefold system was superseded by the Solvac holding, which was set up in January 1983 and subsequently introduced in the stock exchange as a registered quotation alternative to the new nonregistered Solvay shares. The creation of Solvac was a great success; the holding, which was not open to institutional investors, decisively enhanced the exchange of shares and contributed to overcome the threat of a "capital crunch."

At exactly the same time Solvay shares were introduced, another long-term issue was tackled – namely, that of organizational structure. Although Solvay & Cie's multilayered organizational pattern had proved relevant for decades, several problems of overlap and duplication had emerged after the strategy of diversification had been adopted. Some blamed it on the excessive power of national organizations, and others pointed to the dominant role exerted by the General Technical Division (*Direction Générale Technique*). Taking into account the momentum of the legal regime change, Jacques Solvay took the bull by the horns: He decided to turn to a professional business consultant. Obviously, he was not the first to do so. This was the heyday of U.S. consulting firms, which advised companies to "modernize" their organizational chart. By the end of the 1960s, some three-fourths of the largest firms were transformed into diversified, mostly multidivisional, companies.[10] Jacques Solvay recruited Dick Paget from McCormick and Paget

---

[10] Mauro F. Guillen, *Models of Management. Work, Authority, and Organization in a Comparative Perspective*, Chicago-London, University of Chicago Press, 1994, 88.

consulting firm. They were not unknown to each other: Dick Paget was consultant to Allied Chemical, and Jacques Solvay had been a member of Allied's board since 1966. In the following months, Paget's brainstorming resulted in the establishment of the so-called matrix organization, which amounted to a thorough rationalization of Solvay S.A.'s organizational structure. Among several institutional innovations, it reshaped the dynamics of strategic executive tasks formerly carried out by the *gérance* and now managed by an executive committee, the ExCom. In line with this rationale, two long-time Solvay & Cie managers, Albert Bietlot and Edouard Swolfs, were appointed to the top executive level. Interestingly, theirs was the first nonfamily appointment that was not decided by Ernest Solvay himself. This was an undeniable symbol of change, and although these changes were implemented gradually, the speed of the process was strikingly fast in view of Solvay's time scale.

AN AMERICAN COMEBACK

On 14 June 1971, Jacques Solvay succeeded René Boël as chairman of the board and chairman of the ExCom. In his last speech to the General Assembly, Boël made clear to the personnel and shareholders that he was leaving the company "without the slightest concern."[11] He nevertheless pointed out that 1970 had been characterized by an unprecedented rate of increase in wages. This was but a mere starter of the wage-price spiral that took place in the second half of the 1970s that gave employers recurring nightmares.[12] A couple of months after Boël's retirement discourse, U.S. President Richard Nixon announced that the dollar could no longer be exchanged for gold, a decision that hastily provoked the entire collapse of the international monetary system enacted at Bretton Woods shortly after the war. Underlying Nixon's decision was the important military burden of the Vietnam War, as well as the increasing budget deficits of the U.S. federal budget.[13] As a result of this situation, most industrialized countries were left dazed and

---

[11] "Allocution du Comte Boël à l'Assemblée Générale du 14 juin 1971," *Revue du personnel Solvay*, 1971, June–July (4), 7.
[12] Eichengreen, *European Economy*, 267.
[13] Judt, *Postwar*, 454.

Figure 8.3. Under the watchful eye of the founder, still symbolically present on the premises of the company, an engineer works on an analogue calculator at the research and development center in Neder-over-Heembeek, 1969. (Solvay Archives, photograph by Bauters)

confused about how to handle the international monetary system. Together with the war waged in Vietnam, the end of the gold standard contributed to the erosion of U.S. prestige in the Western world, although for different reasons and among different audiences. At the start of the 1970s, the perception of the United States as an undisputed world leader was waning. Yet the U.S. standards of economic growth, mass consumption, and leisure society still charmed many industrialists, especially in Western Europe. The model did not convey the "American dream," as it had in the aftermath of the war, but it nevertheless outlined the idea of the "American challenge," which the French essayist Jean-Jacques Servan-Schreiber had aptly coined in 1967.

One could say that Jacques Solvay was among those industrialists who staunchly wished to take up this challenge. The twists and turns of the international monetary system and the hawkish mindset of the Cold War order had not undermined his fascination for the United States. On the contrary, the American model of economic liberalism had strengthened his aversion to the State-driven political economies of continental Western Europe.[14] Like his father, Jacques Solvay held a degree in civil engineering from the Université Libre de Bruxelles; also like his father, he had started his professional career by gaining experience outside the family business. But whereas Ernest-John Solvay had spent several months at Solvay Process Company (SPC) and at ICI, his son had worked at General Electric's Bloomfield works. The impression that they drew from their respective encounter with the American way of life could not have been more different. Ernest-John was wary of the Americans. Taking into account his later industrial experience with Solvay & Cie's American partners, he had reasons to be on guard. During the First World War, the Hazard brothers had attempted to build an American-only coalition of SPC shareholders against the Solvays. It failed by a narrow margin after a lengthy legal procedure. The worst was yet to come during the "Weberian years" at Allied Chemical (see Chapter 5). Hence, Ernest-John Solvay had grown suspicious of any partnership with an American business, as the aborted Wyandotte operation had made clear. Jacques Solvay did not share his father's reluctance, quite the opposite. He had enjoyed his early stay in the United States, where he met the French native but long-time U.S. resident Marie-Claude Boulin, who subsequently became his wife. In the early 1950s, the couple settled in Europe imbued with American nostalgia. Jacques Solvay was determined to come back regularly to the United States – and, one should add, not merely in his capacity as Allied Chemical's shareholder and member of its board.

Yet, Solvay's industrial comeback in the United States only came to fruition in the early 1970s. After the Wyandotte stalemate, Jacques Solvay had followed the lead of Allied Chemical. The objective was to set up a joint venture in high-density polyethylene, based on Solvay's know-how. Resuming business relations with

---

[14] This section draws extensively on Bertrams, Coupain, and Homburg, *Solvay*, chap. 18.

Figure 8.4. Solvay came back to the United States as an industrial player through the acquisition of the high-density polyethylene business of Celanese in 1974. Claude Loutrel, Jacques Solvay, and Whitson Sadler, shown at the plant in Deer Park, Texas, were the main leaders involved in this American comeback. (Solvay Archives)

Allied was a symbolic gesture in many respects, but it was above all a logical move considering Allied's position in the U.S. chemical market. In the end, however, negotiations failed. Was it a dire prediction for future Solvay-Allied relations? Whatever the reasons, Jacques Solvay decided to pursue the effort, away from Allied. Among the many alternatives that were sought, one opportunity remained constant: purchasing of the polyethylene business of Celanese Corporation America, which included huge production facilities at Deer Park, Texas. The deal was finally sealed in September 1974, followed by almost a year of technical postagreement negotiations. The Solvay subsidiary that resulted from the acquisition was called Soltex Polymer Corporation, in part because the right to use the name Solvay was still owned by Allied Chemical in the United States. He had finally done it. Jacques Solvay had successfully taken up his "American challenge." Instrumental in this U.S. comeback story was the role played by Jacques Solvay's own American network, which he had patiently woven since the mid-1960s. Apart from Claude Loutrel, appointed president and CEO of Soltex, the team included the old "anti-Weber" friend George Murnane from Sullivan & Cromwell law firm, the industrial consultant Dick Paget, and high-profile members of the Lazard Frères & Co. private bank (its head, André Meyer; his successor, Franck Pizzitola; and a promising man named Whitson Sadler). With the same advisers, Jacques Solvay decided to pull Solvay S.A. out of Allied Chemical. From a legal (antitrust) standpoint, the creation of Soltex had made the association of Solvay with Allied tenuous. Solvay thus progressively sold its Allied shares, from 1975 until 1985, when Allied Chemical merged with Signal Companies to become Allied-Signal (taken over by Honeywell in 1999). In some ways, the birth of Soltex helped Solvay to definitively turn the page with Allied.

CAUGHT UP BY THE CRISIS

Happy-ending stories only take place in Hollywood, however, and Deer Park was in Texas, far from California. Dark clouds soon stained Soltex's blue sky. Not surprisingly, the acquisition of a high-density polyethylene had prompted the company to secure feedstock. What was not expected, however, was the far-reaching outcome of the oil crisis, which was in full swing in 1975. The price of ethylene was soaring, and a solution was urgently needed but required substantial investment. It came from two angles. First, in cooperation with ICI and other American firms, Soltex took 25 percent interest in Corpus Christi Petrochemical Company. The firm was established to construct a huge steam cracker and a network of pipelines. Second, management decided to erect a polypropylene plant at Deer Park itself. These constructions came at a rough time because overcapacity was looming. Furthermore, 1980 witnessed a general "plastics crisis."[15] Chemical companies active in the manufacturing of polymers were strongly dependent on rising oil

---

[15] Bertrams, Coupain, and Homburg, *Solvay*, chap. 20.

prices, as well as a series of interlocking technological constraints. For instance, steam crackers, like Corpus Christi's, cannot function below a certain capacity (approximately 60 percent) without suffering major losses.[16]

With its feedstock supplies curtailed, the plastics industry simultaneously faced a growing opposition from ecologically minded activists. Because of the wide array of disposable products it manufactured and its strong reliance on petroleum, the plastics industry was under attack. At approximately the same time, public awareness increased strongly on the possible health dangers of emissions of chlorinated organic substances. Mostly inspired by nongovernmental environmental organizations such as Greenpeace (created in 1971), these protests would reach their heights in the 1980s and 1990s.[17] Environmental issues were thus added to macroeconomic concerns. "It all came together to form a powerful antiplastic mentality that had little impact on legislation or on actual consumption of plastic but that perpetuated a compelling negative image."[18] On the other hand, the problem went beyond one industry in particular. The second oil crisis, which broke out in 1979–80, had terrible consequences on the industrial economy in general, and on the steel and chemical industries in particular. Restructuring and reorganization were taking place on an international scale. In France, however, this led to a singular situation. There, rationalization was immediately followed by a sudden wave of nationalization in the wake of the election of François Mitterrand. Next to the major steel producers (Usinor and Sacilor), the leading French chemical companies were taken over by the State in 1982 (Rhône-Poulenc, Péchiney, and Saint-Gobain).[19]

In this recessionary context, Soltex was at the forefront of the crisis. However, with regard to the global magnitude of the depression and the scope of investments realized in polyolefins product lines, the downturn affected Solvay S.A. as well.[20] In 1981, for the first time in its history, the company announced it had recorded a

---

[16] Aftalion, *International Chemical Industry*, 321.
[17] Bertrams, Coupain, and Homburg, *Solvay*, chap. 21.
[18] Meikle, *American Plastic*, 271.
[19] Michel Margairaz, "Mai 1981," in Jean-Claude Daumas (dir.), *Dictionnaire historique des patrons français*, Paris, Flammarion, 2010, 1343–8.
[20] Bertrams, Coupain, and Homburg, *Solvay*, chap. 20.

net loss. This turned out to be a stunning shock to everyone, for if a difficult year had been foreseen, the degree of the collision had not been fully anticipated within the Solvay group.[21] Not surprisingly, the bulk of the losses were incurred in the polyolefins business. The Corpus Christi cracker was a huge financial burden. It had caused a significant increase of Solvay's traditionally low debt ratio during the 1970s and was now the source of a major financial setback for the group. The focus on Corpus Christi tended to overshadow other domains that had also been hit hard by the slump, notably in Europe. For instance, the production levels of PVC (Solvic) plants were notoriously lagging behind. Solvay's partner in the joint venture, ICI, refused to compensate for the cash shortage and made clear its intention to opt out from Solvic. Despite a difficult cash-flow situation, ICI's withdrawal paved the way for the progressive complete integration of Solvic plants into Solvay, which began in 1984–5.

With profits sagging and prospects for growth limited, Solvay's shareholders became anxious. They demanded radical measures. A series of initiatives had already been taken, although it would be several years before they paid off. At Soltex, technological improvements in the polyethylene and polypropylene sectors resulted in redefining and widening the product lines of the polyolefins business. This alleviated the cyclical character of this vulnerable domain. Because overcapacity was a global issue, Jacques Solvay managed to drive a coordinated action against the crisis on the European level. Of course, European industrialists had to take into account dispositions for securing the principle of free competition that had been enacted in the Treaty of Rome establishing the European Economic Community in 1957. Thanks to his contacts with the EEC commissioner for industrial policy, Viscount Etienne Davignon, who had distinguished himself in 1977–8 with two ambitious plans to boost the European steel industry,[22] Jacques Solvay succeeded in setting out a provisional "crisis cartel" to lower the production levels, which immediately diminished the threat of overcapacity.

---

[21] ACS, "Bilan économique 1968–1988. Essai de diagnostic et conclusions," by Paul Washer, 27 May 1988.
[22] René Leboutte, *Histoire économique et sociale de la construction européenne*, Bern-Brussels, Peter Lang, 2008, 486–91.

On the other hand, next to these transitory responses, Solvay and its competitors resorted to the more traditional defensive means: cost cutting. Obsolete and unprofitable plants were closed down, like the diaphragm electrolysis at Rosignano (1981) and at Jemeppe (1982), the PVC unit at Torrelavega (1983) and the soda ash plant at Sarralbe (1983). Industrial activities that were not indispensable to the core business (synthetic paper pulp, foams, paints and coatings, etc.) were divested while investments were made in new areas. For instance, in the first half of the 1980s, Solvay's processing division moved in, and later focused on, the automobile industry through the production of fuel tanks. At the insistent request of Solvay's newly appointed CEO, Daniel Janssen, the emblematic Corpus Christi cracker was ultimately sold in 1987. Finally, staff-cut programs were implemented in the Belgian historical plants of Couillet and Jemeppe, terminating in "defensive" strike waves that were typical of a context of economic turmoil.[23] These efforts led to a contraction of 10 percent of Solvay's personnel worldwide during the 1980s. Harsh as they were, these measures proved successful. Already by the mid-1980s, annual reports mentioned positive and promising results. Within Solvay's industrial portfolio mix, a recently explored segment showed particularly great potential.

BIOCHEMICAL INNOVATION

As historian Alfred Chandler summarized it, after the second oil crisis, Solvay's "restructuring was implemented in much the same fashion as that of American core companies . . . : buying units closely related to the lines in which its organizational capabilities ensured its competitive strength, and selling off those less related to its successful products."[24] But from what learning bases did Solvay effectively start to refocus its industrial development? Before answering this question, a broader overview is needed. In addition to its devastating consequences, the economic crises of the 1970s also prompted positive transformations, which brought the catchword *innovation* to the forefront of the corporate culture. Most of the new technologies that first appeared stemmed from non–capital-intensive

---

[23] Vinen, *History in Fragments*, 441.
[24] Chandler, *Shaping the Industrial Century*, 136–7.

small and medium businesses, some of which were closely connected to university research laboratories. National governments acted as stimulators for this new breed of technology transfer. In the United States, the Bayh-Dole Act passed in 1980 enabled universities to license patents to companies and collect royalties, even if government grants had initiated the research projects.[25] The impetus came a bit later in the European Economic Community but was no less noteworthy. Recalling this era of new technology, the electronics and computer industry, especially sophisticated in Japan at the time, directly comes to mind. But the area of biotechnology, which grew after a series of joint industry-university breakthroughs in recombinant DNA and genetic engineering techniques, was also perceived as a young, emerging field. In contrast to the computer industry, however, biotechnology did not operate through an accumulation of incremental improvements. It was a radically new application of basic science, a technology in which those who understand an idea "are often scientists and...therefore play leading roles in its introduction."[26] Thus, during the early stages of this revolution, the small science-driven company epitomized the biotechnology industry, but by the 1980s, the picture had changed drastically. A study focusing on the founders of U.S. biotechnological companies showed that "whereas in the period 1971–80 almost twice as many had academic as had business origins, by the mid-1980s, two-thirds were from business."[27]

Like American chemical companies (Du Pont, Dow, Monsanto) and their European competitors (Bayer, Hoechst, Ciba-Geigy, Rhône-Poulenc),[28] Solvay S.A. entered in the life sciences in the late 1970s. The move was made in two successive, but rather distinct, timeframes. The first step took place early in 1975 and was

[25] David C. Mowery, Richard R. Nelson, Bhaven N. Sampat, and Arvids A. Ziedonis, *Ivory Tower and Industrial Innovation: University-Industry Technology Transfer before and after the Bayh-Dole Act in the United States*, Stanford, Stanford University Press, 2004, esp. 85–98.
[26] David C. Mowery and Nathan Rosenberg, *Technology and the Pursuit of Economic Growth*, Cambridge/New York, Cambridge University Press, 1989, 260–1.
[27] Bud, *Uses of Life*, 193.
[28] Hounshell and Smith, *Science and Corporate Strategy*, 589–90; Chandler, *Shaping the Industrial Century*, 283–85.

not the outcome of a deliberate plan of action. It consisted of raising the shareholding on one of Solvay's subsidiaries based in Germany, Kali-Chemie, which was a small but dynamic and diversified chemical multinational company.[29] The story of Solvay's interest in Kali-Chemie dated back from the interwar with DSW holding some 20 percent of the capital. But it was not before the 1950s and the dismantling of the I.G. Farben concern that DSW succeeded in obtaining the majority. The operation, which proved extremely delicate, was orchestrated by the discreet and efficient banker Hermann Abs.[30] Since then, however, Kali-Chemie had been developing its product lines quite independently from Solvay. After an accident occurred in a potassium mine owned by Kali-Chemie in 1975, DSW was able to increase its holding from 65 percent to 78 percent. According to German corporate law, the Solvay group could now orient Kali-Chemie's business strategy, although, for cultural reasons, it took years before an effective integration could be realized. For several years, therefore, Kali-Chemie pursued its policy of acquiring businesses in the pharmaceutical sector. Looking at its turnover and profits, its competitiveness was remarkable. During the struggle to tame the consequences of the oil crisis in Europe and in the United States, Kali-Chemie's performance undoubtedly gave Jacques Solvay some ideas.

The second step was realized in 1977 when the Board of Solvay S.A. finally endorsed a new business strategy. Initial talks had focused on moving toward the manufacturing of noncyclical products. The group keenly sought an alternative to petrochemical feedstocks, which were particularly strained in the United States. It is thus not a coincidence that Solvay's first acquisition in the domain of life sciences took place there. In 1979, Soltex acquired Salsbury Laboratory, a company specialized in animal health and world leader in the production of poultry vaccines. The deal had been prepared by Jacques Solvay himself, with the assistance of his American network. This purchase gave way in 1984 to the transformation of Solvay's former U.S. holding into a new single entity called

---

[29] This section draws largely on Bertrams, Coupain, and Homburg, *Solvay*, chap. 19.
[30] For this episode, see Bertrams, Coupain, and Homburg, *Solvay*, chap. 13.

Figure 8.5. Karol Wojtyła was hired during the Second World War in a quarry, then in a plant owned by Solvay in the Krakow area. Thanks to this, he avoided deportation to Germany. After his election as Pope John Paul II, he paid a visit to the Rosignano plant in 1982. (Solvay Archives)

Solvay America, with Whitson Sadler appointed as CEO. In the following years, Solvay America became the group's entry for expansion in the biochemical sector through the acquisition, spurred by Daniel Janssen in 1986, of the pharmaceutical company Reid Rowell. In the meantime, another major purchase confirmed Solvay's "biostrategy." After more than a year of negotiations, the Dutch pharmaceutical company Philips-Duphar, which had resulted from the curious wedding between a multinational electronics company (Philips) and a chocolate producer (Van Houten), was purchased by Solvay in 1980. Duphar's major industrial activities were pharmaceuticals, veterinary products, vitamins, and fine chemicals, as well as crop protection. The latter product line contributed to sully the image of the Dutch company shortly after dioxin soil pollution was discovered in 1980, roughly four years after the far-reaching dioxin contamination occurring at Seveso in Italy. There were interesting synergies with Kali-Chemie and Salsbury altogether. Whereas the German company's core value was in the sales and marketing of its products, Duphar's focus was on innovation, with almost a quarter of its sales budget spent on research.

After the first wave of branching out into polymer and petrochemical activities in the 1950s, Solvay was a different company from the one that came out after the Second World War. Solvay's second wave of diversification into life sciences went a step further in this development. Looking at the 1988 annual report, the year of the company's 125th anniversary, the outcome of this strategy are stunning. The division "alkalis, chlorine and derivatives" accounted for 30.7 percent of Solvay's turnover, "peroxides" for 7.4 percent, "plastics" for 31.8 percent, "plastics processing" for 18.1 percent, and "health" for 12 percent.[31] For Solvay, the biochemical revolution radically changed the company's profile. From 1988 to 2000, the health sector, with special emphasis on human pharmaceuticals, represented a huge contribution to the company's sales and especially its gross profits. Actually, the total group was active in only three sectors – chemicals, plastics, and pharmaceuticals – the last being the most profitable and fastest-growing and not cyclical. By 2009, more than 60 percent of the group's profits came from pharmaceuticals. Biochemicals had changed Solvay. Another major milestone in the family firm's history was that, despite psychological concerns, it had been transformed into a joint-stock company. In addition, the business had positively reacted to an enduring crisis, which had hit the chemical industry and the oil-consuming sectors particularly hard. Of course, successful reactions to events at a given time may later present unexpected challenges. This proved true as the Cold War drew to an end. Solvay had successfully taken up the challenge of diversification. Now it needed to adapt to an increasingly diversified world.

---

[31] Maxime Rapaille, *Solvay, un géant*, Bruxelles, Didier Hatier, 1989, 135.

Figure 9.1. Since 2004, Solvay has been one of the main sponsors of Bertrand Piccard's and André Borschberg's Solar Impulse project to develop a solar airplane. Solvay contributed extensively to this project through its research efforts, as well as its fluoropolymers and engineering plastics for batteries, photovoltaic cells, and lightweight constructive parts. (© Solar Impulse / Stéphane Gros)

9

Globalization and Consolidation

The oldest of Central European tales has a German general saying to his Austrian ally: *"The situation is serious but not tragic,"* to which the Austrian general replies: *"No, mein Lieber, the situation is tragic but not serious."* One might say that the situation of East Central Europe today is tragic but not hopeless.

Timothy Garton Ash (September 1988)[1]

In the 1980s it was fashionable to argue that a vibrant Japan would overtake the United States and that the United States urgently needed to remake its institutions along Japanese lines. In the first half of the 1990s, when productivity grew faster in Europe, it was argued that the United States needed to remake its economy along European lines... Today both of these examples of "systems envy" have fallen out of fashion. For anyone encountering forceful statements of American triumphalism and Eurosclerosis, history is a reminder that this too shall pass.

Barry Eichengreen (2007)[2]

THE COLD WAR IS OVER

On 2 May 1989, the Hungarian authorities decided to remove the electrified fence on the western side of the country, although the border remained officially closed. In less than a couple of months, some 25,000 East Germans swarmed into the country to take a

---

[1] Timothy Garton Ash, *The Uses of Adversity. Essays on the Fate of Central Europe*, New York, Vintage, 1990, 300.
[2] Eichengreen, *European Economy*, 414.

vacation there. Early in September, a journalist asked the Hungarian foreign minister what the government's reaction would be if these people intended to make their way to Austria, he replied, "We will allow them through without any further ado and I assume that the Austrians will let them in."[3] Immediately, thousands of East Germans rushed westward. This was the first breach in the Iron Curtain and the spark that ignited an endemic panic within East European Communist parties. Two months later, on 9 November 1989, after many demonstrators in GDR called on Soviet leader Mikhail Gorbachev to take responsibility, a news conference took place in Berlin. The government's press spokesman was asked to clarify when the provisions of a new legislation allowing for foreign travel would come into effect. The spokesman had been on vacation and had received no clear instructions from the party on that matter. Although he had no clue whatsoever, he answered, "From this moment."[4] This stunning declaration prompted thousands of people to head for Checkpoint Charlie, the Brandenburg Gate, and other sites along the border with West Berlin. By the time the control guards received further notice, people had gone west. The Berlin Wall was falling apart. The Soviet Union, the seedbed of Communism, would follow suit in 1991, leaving free-market capitalism alone in the ideological ring. For many observers, this marked the conclusion of the "short Twentieth Century."[5]

Taking the collapse of the Soviet empire into account, the unexpected end of the Cold War in 1989–91 did not alter the geopolitical configuration inherited from Yalta and Potsdam agreements of 1945. In Eastern Europe where these borders could have been subject to discontent, two exceptions proved the rule. One was the peaceful "velvet" revolution of 1993, which divided Czechoslovakia into two states; the other was the bloody series of wars in Yugoslavia that reactivated the "Balkan vortex" between 1991 and 1999.[6] On the other hand, it became clearer in hindsight that the fall of the Berlin Wall did not mechanically induce a stronger, united

[3] Judt, *Postwar*, 612–13.
[4] Mazower, *Dark Continent*, 394–5.
[5] Hobsbawm, *Age of Extremes*, 5–11, 484–7.
[6] Misha Glenny, *The Balkans. Nationalism, War, and the Great Powers, 1804–1999*, London, Penguin, 1999, 634–62.

Europe.[7] In the midst of discussions surrounding the so-called reunification of Europe, eyes converged on Germany. Just as in 1919 and 1945, the country was transformed into Europe's political testing room. Contrary to the two former postwar eras, however, the "third postwar" conveyed the prospects of reconciliation and federation. More generally, the post–Cold War era was no longer, or could no longer be, Western-centric. It would be tempting to argue that the world was becoming more multilateral. Yet it appeared that the East-West divide was superseded by an array of other divisions affecting the global sphere, the North-South division being the most enduring. Some thirty years after the ultimate wave of decolonization, the world economy was still unequally balanced, albeit new industrial players had come to the fore. The infatuation with emerging markets – the BRIC (Brazil, Russia, India, and China) acronym was coined in 2001 – has contributed to overlooking and distorting the global picture of modern globalization, which ultimately had much in common with the wave of globalization that started by 1870 (see Chapter 3).[8]

"THE WIND OF CHANGE": BACK TO CENTRAL EASTERN EUROPE

To a large extent, the reunification of the two Germanys was in the logic of history. Yet from the former "Big Four" powers, U.S. President George Bush (senior) was the only one in favor of the process. At various times, Margaret Thatcher, François Mitterrand, and Mikhail Gorbachev had all expressed both their concern and their reluctance to their German counterpart, the Federal Chancellor Helmut Kohl.[9] Nothing of that sort happened at Solvay; hesitation did not take hold of the company's CEO at the time. The grandson of Emmanuel Janssen, Daniel Janssen (b. in 1936), had been asked to join Solvay in 1984 while he was running the chemical and pharmaceutical business Union Chimique Belge set up by his

---

[7] Vinen, *History in Fragments*, 465–9.
[8] Philippe Norel, *L'invention du marché. Une histoire économique de la mondialisation*, Paris, Seuil, 2004, 430–2.
[9] Heinrich A. Winkler, *Germany. The Long Road West, 1933–1990*, vol. 2, Oxford/New York, Oxford University Press, 2000, 519–21.

grandfather in 1928 shortly before he withdrew from Solvay. Appointed CEO in 1986, he knew too well the dramatic history of the loss of Bernburg as told and experienced by his uncle René Boël. But there was more to the story than historical bindings. Before the fall of the Berlin Wall, Daniel Janssen had initiated an important transformation of Solvay's activities in Germany. The objective was to integrate Solvay's subsidiaries, DSW and Kali-Chemie, into a new entity of the Solvay group.[10] The process led to the creation of Solvay Deutschland and was in motion when the Berlin Wall fell.

Contrary to many East European countries, the GDR authorities never paid Solvay a pfennig in compensation for the "supervision" of Bernburg and the five other plants located in East Germany. The official recognition of the GDR by the Belgian government in 1972 had given leeway in the negotiations, but it proved short-lived. As a result, Solvay demanded to be allowed to retrieve Bernburg free of charge. When discussions resumed by March 1990, diplomatic and industrial negotiations overlapped. On a political level, there were still many legal obstacles to overcome. A monetary and economic unification agreement between both Germanys was completed in July 1990, while the unification treaty came into force on 3 October of the same year.[11] At the level of the factory, things were no less difficult. For the Solvay delegation coming from Brussels, the obsolescence of the Bernburg facilities was striking. The soda plant still employed 1,800 workers and was running in overcapacity. Obviously, the economics of the central planning system had gone wrong. A final agreement came into effect on 1 September 1991 that foresaw an important reduction of the workforce and a modernization plan of the main facilities. However, partly due to the hasty mechanisms of reunification, the economic situation of Germany rapidly deteriorated. The early 1990s were rough for Solvay as well, with results below expectations. Cost-cutting schemes were implemented everywhere in Europe (as discussed later in the chapter). The integration of Solvay Deutschland thus took place at a difficult moment. Ultimately, however, efforts and substantial investments

---

[10] This section draws largely on Bertrams, Coupain, and Homburg, *Solvay*, chap. 20.
[11] Eichengreen, *European Economy*, 318–19.

Globalization and Consolidation

Figure 9.2. The Bernburg plant (here in 1991) had been sequestered by the East German Communist regime. It was recovered after the fall of the Berlin wall for one symbolic German mark. (Solvay Archives)

paid off: The new Bernburg soda ash plant came into being in 1994, followed by the hydrogen peroxide unit in 1995.

The ink was barely dry on the Bernburg agreement when Solvay decided to push farther east in Europe. Of course, remembering the "good old days" of Solvay-Werke, Zaklady Solvay, and other fruitful combinations in Central Eastern Europe, the company had many reasons to do so. But by 1990s standards, the region had become a *terra incognita*. Quite expectedly, the first breakthrough came with trade and commercial contracts. Even in the days of the GDR, many "Solvay products" had circulated east of the Iron Curtain. Now official sales offices had to be set up in the region. Before talking about industrial expansion, the shift from state socialism to free-market capitalism had to be achieved. The "transition," if there was one, was harsh: Central and Eastern Europe were hit by a severe crisis. Some economists talked about necessary structural "readjustments"; others pointed to misleading decisions in terms of political economy. As one observer put it, "a regulated market instead of a self-regulating market, a mixed economy with a restructured and efficient state-owned sector, at least initially, would have generated a more organic transition from plan to market. Such an approach, however, was immediately rejected."[12] As a result, in the three first years of the 1990s, GDP and output across the region declined between 25 and 30 percent and 30 and 40 percent, respectively. In this context, negotiations dragged on for years. Only in 1996 could Solvay announce the conclusion of an agreement – namely, the acquisition of Devnya soda ash plant in Bulgaria, near the Black Sea.[13] This was a major achievement. Devnya was a modern plant and the largest single-unit plant in Europe in terms of capacity. In the following years, thanks to large investments, the newly baptized Solvay Sodi company became the bridgehead for Solvay's expansion in markets located in southeastern Europe and into the Middle East and Central Asia. Ironically, together with Albania, Bulgaria was the only country where Solvay had no factory before 1945. The picture in the 1990s was thus reversed as Solvay's attempts to regain control of its former plants in Poland (Podgórze and Inowrocław) and in Russia (Berezniki) had failed.

TIGERS AND DRAGONS: SOLVAY IN ASIA

The end of the Cold War era surely transformed European societies, but another process was unfolding at the same time, a process that was broader and more profound: the start of the second "fin-de-siècle" globalization. Indeed, it became overwhelmingly clear that there was a world beyond, rather than behind, Europe or the United States. From about the mid-1980s until a sudden and harsh financial crisis struck in 1997–8, the so-called Pacific Asian region was the fastest-growing territory in the world economy. It concerned an arc of countries from Japan and Korea in the northeast to

---

[12] Berend, *Economic History*, 186.
[13] Bertrams, Coupain, and Homburg, *Solvay*, chap. 21.

Figure 9.3. In the early 1990s, Solvay returned to Central Eastern Europe. In 1996, the company acquired the large soda ash plant at Devnya, Bulgaria, on the shore of the Black Sea. This plant was the pride of the country: It appeared on ten-leva Bulgarian banknotes at the time. (National Bank of Bulgaria)

Indonesia, Singapore, and southern China in the southeast. Ample evidence shows that these economies were growing at 8 percent a year and accounted for approximately two-thirds of world capital spending. Their part in the world output was the most impressive. Between 1980 and 2005, the Asian economy grew from 23.5 percent of the world industrial output to 40.1 percent, whereas European countries accounted for 40.1 percent and decreased to a mere 27.7 percent in the same period. The evolutions of both continents were thus at polar opposites. If we take Southeast Asia in particular, industrial growth increased fourfold in the same twenty-five years.[14] As an economist noted, the spectacular economic growth of the emerging markets of East Asia "transfixed the rest of the world." Before the 1997–8 financial crisis, he went on, "It seemed possible and even probable that these economies would become the center

[14] Laurent Carroué, *Géographie de la mondialisation*, Paris, Armand Colin, 2007, 163.

of the world economy early in the twentieth-first century."[15] Only a handful of analysts dampened the enthusiasm for the Asian growth model. Writing a few weeks before the Asian depression, economist Paul Krugman observed that "barring a catastrophic political upheaval, it is likely that growth in East Asia will continue to outpace growth in the West for the next decade and beyond. But it will not do so at the pace of the recent years. From the perspective of the year 2010, current projections of Asian supremacy extrapolated from recent trends may well look almost as silly as 1960s-vintage forecasts of Soviet industrial supremacy did from the perspective of the Brezhnev years."[16]

After modest commercial initiatives in the 1960s and 1970s, Solvay finally launched a far-reaching industrial offensive in Asia in the early 1980s.[17] The first steps had been taken by Jacques Solvay's close collaborator and ExCom member, Claude Loutrel. As soon as he became CEO in 1986, Daniel Janssen gave the Asian strategy a strong and decisive impetus. A first mission in Southeast Asia had been organized in January–February 1985, which resulted in the opening of sales offices in Singapore and Hong Kong, and later in South Korea and Japan as well. Other trips soon followed suit. Several possibilities to initiate industrial activities were explored, with locations in China, Indonesia, and Taiwan. Finally, Thailand was chosen in 1986. It was decided that a hydrogen peroxide plant would be erected via the Interox (Solvay-Laporte) joint venture at Map Ta Phut, south of Bangkok, with easy access to natural gas in the Gulf of Thailand. Peroxythai, as it was called, rapidly gave way to another ambitious project, namely the construction in 1988 of a PVC plant, Vinythai, again at Map Ta Phut. It would be considerably enlarged in the first decade of the twenty-first century. At approximately the same time, the South Korean destination was gaining momentum. A series of agreements were signed, first between Kali-Chemie and Samsung for the manufacturing of special

---

[15] Robert Gilpin, *The Challenge of Global Capitalism. The World Economy in the 21st Century*, Princeton and Oxford, Princeton University Press, 2000, 265.
[16] Paul Krugman, *Pop Internationalism*, Cambridge, MA, MIT Press, 1997, 184.
[17] Bertrams, Coupain, and Homburg, *Solvay*, chap. 20.

Figure 9.4. Solvay responded to globalization by increasing its foothold in Asia. In 1988, on the occasion of the 125th anniversary of the company, Solvay representatives started a "charm offensive" in Southeast Asia to enlarge their network. CEO Daniel Janssen (in the middle) is shown here meeting King Bhumibol of Thailand (on the right), together with Belgian Ambassador Patrick Nothomb (left). (Solvay Archives)

salts needed for the production of television screens, then between Solvay and Nissan's subsidiary, Kasai, in the plastics processing division. Finally, the Japanese market was accessed in a two-step move: through the creation of a subsidiary called Solvay Development (Japan) in 1985 and with the launching of the Nippon Solvay sales office in Tokyo in July 1987.

For Solvay, Thailand, South Korea, and Japan were merely the beginning. Negotiations had started in India as well. With the exception of contracts signed in animal health activities that were divested in 1996, however, they proved limited in the first stage. Only in 2006 did Solvay acquire Gharda Polymers and in May 2012 officially confirm the opening of a new R&D center in Savli, in the state of Gujarat. This largely confirmed that the attraction for India's development did not pale through the years. In addition, back in the late 1980s, Solvay had begun considering the huge Chinese market. Echoing Singapore's long-time prime minister, Lee Kuan Yew, who had attracted the world's most powerful corporations to forge a new kind of state capitalism on the island, Chinese leader Deng Xiaoping had since 1978 been pursuing a policy of economic openness that embraced globalization together with corporatism and authoritarianism.[18] Compared with his rival Mao, Deng was a business pragmatist, as can be illustrated by the famous saying he is credited with: "It does not matter if a cat is black or white, so long as it catches mice." Notwithstanding the welcoming spirit, doing business in China required foreign entrepreneurs to display more diplomatic than hunting skills. It took years for Solvay to succeed in establishing a fruitful business in China. But once the initial contacts had been made, the prospects were promising.

AFTER 1993, FOCUSING ON THE CORE BUSINESS

The euphoria of deregulation and rapid growth in the late 1980s was short-lived. The outbreak of the first Gulf War ushered in a period of financial and economic turmoil, which peaked in the years 1990–3. From the mid-1990s into the first decade of the twenty-first century, periods of (sometimes intense) growth coexisted on a global scale with the financial and monetary slumps that were experienced in the following countries and regions: Mexico (1995), Asia (1997–8), Brazil (1998), Russia (1999), Turkey and Argentina (2000), Turkey, Brazil, and Uruguay (2002).[19] It is too early to claim that these episodes were part of a chain reaction that ultimately led to the world recession of 2008–9. What is sure, however, is that economies had, and still have, to cope with the haphazard movements of the world's financial system. In addition to stock market difficulties,

---

[18] "State Capitalism: The Visible Hand," *The Economist*, 21 January 2012.
[19] Cécile Bastidon-Gilles, Jacques Brasseul et Philippe Gilles, *Histoire de la globalisation financière*, Paris, A. Collin, 2010, 234–72.

international political crises such as the terrorist attacks of September 11, 2001, in New York City and Washington, DC, did not contribute to stabilizing the planet. On the other hand, the first decade and a half or so of the new century has witnessed a sharp increase of global competition. The prospects of reduced labor cost and other social and economic differentials between regions gave way to a phenomenon of industrial relocation in the developing world. Conversely, it enhanced the ongoing movement of deindustrialization in traditionally "advanced" economies through the closure of factories and the disappearance of the working class. During U.S. President Bill Clinton's administration, international agreements and supranational organizations somehow attempted to harness the process of international trade and financial liberalization, but without much success.[20]

Solvay did not escape the recession of 1991–3. Just like the oil crises of the 1970s, the industry of bulk plastics was struck severely by the crisis.[21] In contrast to previous events, however, the division of alkalis could not offset the losses in plastics for long, and it was in the red in 1993. With net growth plummeting, 1993 was Solvay's worst year ever. This prompted Daniel Janssen to launch a worldwide plan of restructuration. It included shutting down installations at Jemeppe, Tavaux, and Linne-Herten, as well as closing down the PVC plant at Hallein in Austria, the soda ash factory at Heilbronn in Germany and, painful as it was for Ernest Solvay's heir, the historical soda ash plant at Couillet, where it all began. The idea had been in the air for some time, but it had always been postponed "until further notice." The decision was finally announced on 24 May 1993. It was a devastating shock for the personnel and for the region as a whole. Solvay's social and welfare legacy had always been praised. Janssen admitted later that "this was the most painful decision of [his] career." If one takes into account the spectacular public campaigns organized in

---

[20] Ian Tyrell, *Transnational Nation. United States History in Global Perspective since 1789*, Basingstoke/New York, Palgrave MacMillan, 2007, 220-1.
[21] This section draws extensively on Bertrams, Coupain, and Homburg, *Solvay*, chap. 21.

March 1993 by Greenpeace and other ecological activists against Solvay's manufacturing activities in chlorine chemistry in the wake of the 1992 United Nations environmental conference in Rio de Janeiro, it is fair to say that 1993 was Solvay's *annus horribilis*. However, to restore profitability and reverse the structural tide of recession, the ExCom decided to carry on the plan for restructuration after the first glimpse of recovery was perceptible the very next year. As a result of gradual downsizing that took place between 1992 and 1998, the workforce was reduced from 46,858 to 33,100 persons.

In parallel with the implementation of its social plan, Solvay centered its strategy on three objectives: strengthening managerial and product leadership, increasing profitability, and expanding its core business. In varying degrees, this was in line with the policy followed by other global chemical companies at the same time.[22] After the 1970s drive for product diversification and the mid-1980s thrust toward geographic spread, the mid-1990s set the stage to a refocusing on the company's main lines of strength. Evidently, the rising weight of global competition had contributed to slim down its portfolio in product lines, learning bases, and corporate culture, through which Solvay had gained competitive advantages through the years. But the instability of the economic situation as well as the deterioration of the image of chemical business due to industrial accidents and environmental impacts were factors that called for a change in practice and culture. It is not a coincidence, therefore, that Solvay S.A., like other chemical companies, started to integrate the principles of sustainable development, responsible care, and risk communication. More generally, the ideas of business ethics, social betterment, and environmental accountability encompassed in the all-embracing patterns of "corporate social responsibility" took hold of large companies. On 24 June 1995, an influential newspaper op-ed set the tone: "Tomorrow's successful company can no longer afford to be a faceless institution that does nothing more than sell the right product at the right price. It will have to present itself more as if it were... an intelligent actor, of upright character, that brings explicit moral judgments to

---

[22] Abelshauser et al., *BASF: The History of a Company*, 616–17.

bear on its dealings with its own employees and with the wider world."[23]

At Solvay, the focus on core and leadership products naturally concerned soda ash. Throughout the 1990s, three large-scale soda ash plants were acquired: at Bernburg in Germany (1991), at Green River in the United States (1992), and at Devnya in Bulgaria (1997). The acquisition of Green River plant of Tenneco, which provided natural soda ash, can be seen as the peak of Solvay's efforts to obtain a strong and undisputed foothold in soda ash in the United States. By doing this, Daniel Janssen completed a project initiated by René Boël and continued by Jacques Solvay. Despite the high price, Green River improved the company's global position by reducing the import pressure from Europe and opening the road to Asia and Latin America. In terms of sustainable development, the acquisition proved to be a visionary decision.

Acquiring the Devnya plant also proved a strategic and timely decision. Taking over this huge soda ash plant involved facing an utterly competitive arena of buyers and dealing with the risk of the unstable economic situation in Bulgaria. This did not weaken Solvay's motivation, and the final agreement was reached in April 1997. As we know, PVC has been Solvay's other flagship product since the early 1950s. Thanks to its plants in Thailand and in South America, where Solvay reinforced its position, the company was a world competitor in PVC manufacturing. The threat of overcapacity in the late 1980s led the company to reshuffle cards. Basically, it amounted to favor synergies for its feedstock strategy. An important merger agreement was thus concluded in 1999 with BASF, in which Solvay could have access to crackers for the supply of ethylene and other raw material. The deal, which allowed Solvay and BASF to concentrate their PVC production on fewer and bigger units, was followed by the closing of plants (Ferrara and Lillo-Antwerp) and other partnership agreements with major oil companies in France

---

[23] *The Economist*, 24 June 1995, quoted in Kevin T. Jackson, *Building Reputational Capital*, Oxford/New York, Oxford University Press, 2004, 19. In 2012, the latter author held the chair of the Daniel Janssen Professorship in Corporate Social Responsibility at Solvay Brussels School of Economics and Management, Université Libre de Bruxelles.

and Spain. This round of measures enabled Solvay to move with confidence on the road toward the third millennium.

LOOKING FORWARD IN THE TWENTY-FIRST CENTURY

In 1998, for the first time in the company's history, a non–family member became Solvay's executive leader.[24] New CEO Aloïs Michielsen (b. in 1942), who succeeded Daniel Janssen, now appointed chairman of the board, was not unknown to the company. As a matter of fact, he perhaps knew more of the family business than many members of the Solvay dynasty. A chemical engineer from the University of Louvain who participated in a program in business administration at the University of Chicago, Michielsen began working for Solvay in 1969. He worked extensively in the field of plastics processing, a division that has been subject of many transformations since the 1970s. The international rationalization of the polyolefins business was even more impressive after the crisis of 1991–3. A wave of worldwide mergers, acquisitions, and alliances swept the polyethylene and propylene industry. Solvay increasingly felt the shadow of these multinational oil colossuses at its back. It not only affected its global position in the polyolefins business itself, it also threatened its downstream activities in bulk plastics and plastics processing. A consulting company advised a strategic alliance with one of these fully integrated oil corporations. As a result of interpersonal proximity, negotiations with the Belgian oil company Petrofina reached an advanced stage. In 1998, however, the Belgian oil refiner was absorbed by the French Total, paving the way to the birth of TotalFinaElf in 2000, the world's fifth largest oil company. At the same time, British Petroleum (BP) took over the American Amoco in what became the world's largest industrial merger.[25]

Owning a polyolefins plant in the United States ranked among BP Amoco's top priorities. After the latter combine made clear its

---

[24] This section draws on Bertrams, Coupain, and Homburg, *Solvay*, chap. 22.
[25] Francisco Parra, *Oil Politics. A Modern History of Petroleum*, Basingstoke/New York, Palgrave MacMillan, 2004, 324–6.

Figure 9.5. The negotiators of the Solvay-BP deal, through which Solvay exchanged its low value added polyolefins activity for BP's specialty polymers in 2001. In the first row, from left to right: Henri Lefèbvre (Solvay plastics), Michael Buzzacott (petrochemicals BP), Byron Elmer Grote (executive director BP), and Aloïs Michielsen (CEO Solvay). (Solvay Archives)

interest in purchasing Solvay's Deer Park high-density polyethylene plant, Solvay made a counterproposal: BP could progressively take over Solvay's entire polyolefins activities, including the less lucrative European plants. BP, on the other hand, would later sell to Solvay its "Engineering Polymers," which had a strong value added. The negotiations lagged for months but ultimately succeeded in August 2001, shortly before the 9/11 terrorists attacks, which were devastating for the oil business. Through this significant portfolio swap, Solvay killed three birds with one stone. It withdrew from an industry it had sought to divest for some time, it kept a strong position in the United States where BP Amoco's "Engineering Polymers" were concentrated, and finally the advanced special polymer business it obtained was by far less cyclic than the industry it had exited. But there was more. While negotiations with BP were stalling, Solvay had started talks with the Italian company Ausimont, a subsidiary of Montedison that specialized in the production of fluorinated polymers. If the synergy with Solvay was interesting, the complementarities with Solvay after its portfolio swap with BP Amoco were a true blessing. The move proved rather difficult, however. It required the intermediation of an array of private and professional networks. When it finally paid off by December 2001, it was Solvay's largest purchase in its history (compensated through the BP Amoco trade-off). Not long after the acquisition of Ausimont was finalized, Solvay ranked high in the production of highly sophisticated engineering polymers. In a relatively short span of time, the company had once again transformed its industrial profile.

When Christian Jourquin (b. in 1948) succeeded Aloïs Michielsen as CEO in May 2006, he picked up the latter's reins of ongoing transformation. Jourquin's appointment was not really a surprise – he had been at Solvay for thirty-five years (and even more if one takes into account his education at the Solvay Business School of the Université Libre de Bruxelles). Nevertheless, the board's decision was another major breach in the company's tradition: For the first time in 143 years, both top positions in the company were occupied by non–family members. If the family business wanted to prove it was not averse to change, this was strong evidence indeed. As soon as he was in charge, Jourquin capitalized on Solvay's global commitment to innovation initiated by Daniel Janssen and broadened by Aloïs Michielsen. An original illustration of this campaign for innovation was the large sponsorship Solvay granted to the Solar Impulse project starting in 2003. Spurred by the Swiss balloonist and former psychiatrist Bertrand Piccard and the engineer and entrepreneur André Borschberg, the Solar Impulse project aimed to construct an airplane powered exclusively by solar energy. Solvay's involvement went beyond funding; it consisted of providing technical expertise as well as in using a wide array of Solvay products based on its recent high-tech chemicals. As it happened, Bertrand Piccard is the grandson of the physicist and tireless inventor Auguste Piccard, a professor at the Université Libre de Bruxelles, and a regular participant to the Solvay Conferences on Physics. After the first successful tests of a prototype aircraft, the Solar Impulse piloted by Borschberg completed a twenty-six-hour flight on 8 July 2010. This was a huge success for the project and welcome publicity for Solvay. Some setbacks had been reported, however: Borschberg's iPod batteries died because of short freezing conditions.[26]

Meanwhile, a huge business operation was underway at Solvay. A specialist in pharmaceuticals, the division in which he had started his career, Jourquin knew better than anyone the changing conditions of the pharmaceutical business. Just like the petroleum industry, it was affected during the 1990s by a series of mergers and acquisitions that led to the creation of huge research- and capital-intensive companies. At first, in 1994–6, Solvay had reacted by divesting its activities in crop protection, enzymes, and animal health. The focus was on human health only, which generated more than half of R&D budget spending. The strategy paid off at the turn of the twenty-first century with important profits obtained through a strategy of exogenous acquisitions. The most important purchase was the French pharmaceutical company Fournier in August 2005. Long-term considerations were less shining, however. Solvay had to come

---

[26] *The New York Times*, 9 July 2010.

to terms with a series of new drug registration failures after successive trials had been requested by the powerful U.S. Food and Drug Administration. In this context, Solvay's research pipeline was drying out. In addition, giant companies were becoming still bigger with stunning mergers making headlines in the financial press (GlaxoSmithKline in 2000, Sanofi-Aventis in 2004, Pfizer-Wyeth in 2009, and Merck-Schering-Plough in 2009). Solvay Pharmaceuticals, albeit thriving, could barely survive in this ultracompetitive environment. Not unlike Philips-Duphar in the early 1970s, Solvay in the early 2000s was mutatis mutandis lacking critical mass. Options were explored, and discreet investigations were carried out. In 2009, word was out in the well-informed milieus that Solvay intended to sell its pharmaceutical activities. It did not take long to learn that the American company Abbott, the world's eighth largest pharmaceutical company (2012), made the highest bid among several serious candidates. "With this transaction," said Christian Jourquin, "Solvay Pharmaceuticals has found a new strong home, within a respected company with a solid and committed position in the industry."[27] In the following weeks, 9,000 employees were transferred to Abbott. Then, just a year after the acquisition, Abbott announced its plan to reduce its research personnel. Eight hundred former Solvay employees in Weesp and Hannover were concerned.

As Ernst Homburg put it, "Solvay had become a totally different company, standing on two legs – chemicals (44 percent) and plastics (56 percent) – instead of three."[28] As a result, the product of the Abbott deal, no less than €4.5 billion, was subject to many discussions and rumors. The sudden outbreak of the U.S. financial crisis in 2008, with its rapid worldwide extension in the economic sphere, prompted Jourquin not to make a hasty decision. By the end of 2010, the pressure grew considerably regarding an eventual acquisition. A strategy and social plan titled the Horizon Project had been launched at the start of that year with the guidance a consulting company. It advised the introduction of reorganizational measures closer to the costumers and a new round of personnel reduction,

---

[27] *The Financial Times*, 28 September 2009.
[28] Bertrams, Coupain, and Homburg, *Solvay*, 551.

Figure 9.6. The acquisition of Rhodia by Solvay in 2011 was in fact a merger of two groups of similar size. Jean-Pierre Clamadieu, Rhodia's CEO, was called to succeed Christian Jourquin in 2012 at the head of Solvay. Together, they initiated a world tour of Solvay and Rhodia sites to explain the integration logic and answer questions from employees of the new group. (Solvay Archives)

which had to be completed in the next three years. At the same time, the administrative headquarters would move from the historical setting of Ixelles to the Research and Technology Center at Neder-over-Heembeek in north Brussels. On 4 April 2011, Solvay finally announced that a friendly takeover of the French chemical company Rhodia had been agreed to by both parties. It was a major deal in many ways, almost doubling Solvay's annual sales. An offspring of the diversified chemical business Rhône-Poulenc, itself the outcome of a long history,[29] Rhodia became an independent firm in 1999 when Rhône-Poulenc merged its pharmaceuticals activities with Hoechst and Aventis. Rhodia, as a result, specialized in

---

[29] See Pierre Cayez, *Rhône-Poulenc, 1895–1975. Contribution à l'étude d'un groupe industriel*, Paris, Armand Collin-Masson, 1988; *Innovating for Life: Rhône-Poulenc, 1895–1995*, Paris, Albin Michel, 1995.

Figure 9.7. Nicolas Boël (at the forefront) was appointed Chairman of the Board of Solvay in 2012. The grandson of René Boël and Yvonne Solvay, he embodies the continuity of the family's presence at the company's top decision-making echelons. Here, he opens a trona mine methane recovery facility on the Green River, Wyoming site. (Solvay Archives)

chemicals (phosphates), fibers, and polymers. When Jean-Pierre Clamadieu was appointed CEO of Rhodia in 2003, his priority was to increase the focus on core industrial positions and reorganize the company accordingly. In addition to the product complementarities between the companies, Solvay could rely on a favorable geographical synergy. Asia and Latin America, in particular, were the cornerstone of this global expansion.

Quite naturally, a merger of such scope raises the question of integration, which is certainly one of Solvay's most important challenges at its 150th anniversary. A token of the changes to come is already apparent through the designation of Jean-Pierre Clamadieu as successor to Christian Jourquin as CEO of the company as of 10 May 2012. On the other hand, Nicolas Boël, a direct heir of the founding families, replaced Aloïs Michielsen as chairman of the board. In terms of business orientation, the takeover of Rhodia was in line with the strategy focusing on profitability and core leadership products, which had been crafted in the early years of the 1990s. More generally, however, the upcoming managerial tandem epitomized two salient characteristics of the company since its early days: a strong international position and a corporate culture that rested on family leadership.